ID0409301

WITHDRAWN

City of Limerick

Vocational Education Committee

O.P.W. Course. 1964.

Director of Limerick
Institute of Technology

Joe

Hova

RTC Limerick

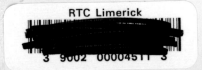

3 9002 00004511 3

Principles
and Use of
Surveying Instruments

BLACKIE & SON LIMITED
16/18 William IV Street, Charing Cross, LONDON, W.C.2
17 Stanhope Street, GLASGOW

BLACKIE & SON (INDIA) LIMITED
103/5 Fort Street, BOMBAY

BLACKIE & SON (CANADA) LIMITED
TORONTO

SECOND EDITION

Principles
and Use of
Surveying Instruments

by J. CLENDINNING

O.B.E., B.Sc.(Eng.)

Surveyor-General, Gold Coast, 1926–1938

London · BLACKIE AND SON LIMITED · Glasgow

©
J. Clendinning

First published 1950
Second Edition, 1959
Reprinted 1960

School of
Engineering
Limerick Technical College
Class. No. 526·9078
CLE
Acc.
No. 288

Printed in Great Britain by Blackie & Son, Ltd., Glasgow

PREFACE

This volume is one of two in which it is intended to give in a limited compass a sound grounding in those parts of the theory and practice of plane surveying that are most commonly used by Civil Engineers and are required by students taking the examination in surveying for the Associate Membership of the Institution of Civil Engineers. The present volume, as its name implies, deals with the Principles and Use of Surveying Instruments, while the second volume deals with the Principles and Practice of Surveying. The two volumes together offer a comprehensive course in the elements of ordinary plane surveying, but it will probably be a convenience to many to have the parts dealing with instruments separate from those dealing with the principles and practice of the main subject. Every effort has been made to limit the text to the minimum necessary to cover the syllabus of the examination, and in so doing it is felt that the ground covered is in fact that which is really essential to a young engineer commencing practice. For this reason, geodesy and field astronomy have been excluded as being outside the scope of the books. The latter therefore do not claim to be a complete treatise on surveying but are to be regarded as a textbook of an elementary or intermediate standard of difficulty.

In the present volume, a good deal of space has been given to instrumental adjustments, as a thorough knowledge of, and practice in, these adjustments is a most necessary accomplishment for any surveyor. For the student, however, no amount of description or reading is an adequate substitute for the actual handling of instruments and he is strongly advised to practice the adjustments by himself whenever he gets the chance. There is nothing inherently difficult in them, and no special dexterity or skill with hand or tool is required, but practice gives confidence, accuracy and speed. Once the adjustments have been mastered, the actual use of the instruments is easy.

It is impossible in a book of this size to describe in detail every instrument that is in use in survey work, but the ones chosen for description are those which are in common use by engineers or are typical of their class.

In conclusion, I should like to express my thanks to Messrs. Cooke,

Troughton & Simms, Ltd.; Messrs. E. R. Watts & Son, Ltd. (now Messrs. Hilger and Watts Ltd.); Messrs. James Chesterman & Co., Ltd.; Messrs. W. F. Stanley & Co., Ltd.; Messrs. Hall Bros. (Optical), Ltd.; Messrs. The Williamson Manufacturing Co., Ltd.; Messrs. Holmes Bros. (London), Ltd., and Messrs. W. Ottway & Co., Ltd., for their assistance in allowing me to use certain illustrations from their catalogues, as noted in the acknowledgement under each illustration, and in some cases for the loan of the blocks or the provision of special photographs.

PREFACE TO THE SECOND EDITION

Apart from some minor corrections, the only amendments of any consequence that have been made in the main text of this book are in the chapter on Tacheometry and in the section dealing with the formulae connecting atmospheric pressure with elevation of station above sea-level in Chapter V, where some matter has been added to take into account some recent work on the temperature correction in the lapse-rate formula and the application of the formula to the graduation of aneroid barometers, the proof of the basic lapse-rate formula now being relegated to the Appendix. At the same time, however, I have taken this opportunity to add an extra chapter to include descriptions of some new self-aligning levels that promise to replace to a considerable extent some of the older types of conventional level and also an account of wedge telemeters and split-image tacheometers which appear to be coming more and more into use in Commonwealth countries and which have added greatly to the accuracy with which lengths may be determined by simple optical means. The chapter also includes a reference to electronic methods of measuring distances, some of which are already beginning to replace the use of invar and steel tapes in trigonometrical base measurement, while others enable very long distances of anything up to about 400 miles or so to be measured relatively accurately.

In conclusion, I wish to thank Colonel D. R. Crone, C.I.E., O.B.E., for information regarding the application of the lapse rate formula as a basis for graduating aneroids and Messrs. Hilger & Watts, Ltd., and Messrs. Wild Heerbrugg Ltd., for providing me with information regarding instruments manufactured by them and for permission to reproduce illustrations from some of their publications.

J. CLENDINNING.

ANGMERING-ON-SEA,
SUSSEX,
29th *October*, 1958.

CONTENTS

CONTENTS ix

Chap. Page

CHAPTER I

INTRODUCTION

In this volume we shall consider the principles and use of the more important equipment and instruments used in ordinary plane surveying (i.e. in surveys where the area involved is so comparatively small that the spheroidal shape of the earth can be neglected and the survey treated as being on a plane surface), leaving the theory and practice of surveying as a whole to be treated in the companion volume on the Principles of Surveying.

The instruments commonly used by the ordinary engineer or surveyor may, for convenience of treatment, be divided into the following main classes:

1. Miscellaneous equipment for ranging and laying out survey lines.
2. Instruments for the direct measurement of lengths and distances. This class includes the ordinary surveyor's chain and steel tapes and bands.
3. Instruments for the measurement of bearings and angles, including the magnetic compass, sextant, theodolite, and miner's dial.
4. Instruments for measuring slopes, elevations and heights. This class includes various kinds of clinometers, hand levels, the Abney level, the well-known surveyor's level, barometers, aneroids, and the hypsometer.
5. Instruments for the measurement of distances by optical means. This class includes tacheometers, subtense bars and range-finders.
6. Instruments for surveying by graphical or photographic means, including the cavalry sketching board, plane-table and photo-theodolite.

The student will find that most surveying instruments require certain adjustments from time to time, and it is essential that he should not only learn how to use and care for his instruments but also that he should be able to test whether or not they are in adjustment, and, when necessary, to make the proper adjustments himself. These adjustments consist in making or maintaining certain geometrical relationships between real or imaginary axes, lines or planes which may be considered to be fundamental concepts in the design of the

1

instrument. In general, when only ordinary standards of accuracy are required, this is not difficult to do, because makers allow for the possibility of instruments getting out of perfect adjustment and make provision in design and construction so that all necessary adjustments can be made reasonably quickly and easily in the field, or in places far removed from all workshop facilities. In some cases, however, where certain adjustments are not likely to be disturbed, they are made permanently by the maker and the surveyor has no control over them. An example of this is where a prism is used to reflect rays of light in particular directions and the faces of the prism are cut to make certain pre-determined angles with one another. Here the user of the instrument is entirely dependent on the prism being correctly shaped.

In many cases it is possible to arrange the methods of observing in such a way as completely to eliminate small faults in instrumental adjustment. A case in point which often occurs in ordinary practice is the prolongation of a straight line on the ground by use of the theodolite. If the " horizontal collimation " of the instrument is at all out of adjustment, and no special precautions are taken in observing, the line set out will be a series of zigzag lines, or else a series of chords of a curve instead of a straight line. If, however, the instrument is used in a particular way and a special procedure is adopted in observing, the effects of imperfections of instrumental adjustment can be neutralized and a true straight line set out with very little difficulty, even when the error in adjustment of the instrument is fairly large. Thus it happens that, in some classes of precise work, notably those which involve observations with a theodolite reading to a second of arc, the surveyor never attempts to secure perfection in adjustment but merely puts his instrument into approximate adjustment and then depends on his methods of observing eliminating the effects of faulty adjustment. In other cases, the numerical values of the errors of adjustment are determined by observation, and corrections for these errors are introduced into the computations of the results. A very simple case of this occurs when the lengths of lines are measured with steel bands that are either too long or too short by small amounts when compared with their nominal length. Here the error in the tape length is found by comparison with a standard tape, or with a standard of length laid out on the ground, and all lengths measured with the tape are corrected to allow for the error.

Instrumental adjustments may be divided into two main classes —temporary and permanent. Temporary adjustments are those which have to be made every time the instrument is set up, while permanent adjustments are those which, once made, can be relied on to remain fixed for a considerable time and do not have to be repeated every time the instrument is used. Thus, when a theodolite is set up over a station or mark, it has to be properly centred, levelled and focused. Centring, levelling and focusing are therefore temporary adjustments which have to be repeated at every set-up. The adjustments for collimation, on the other hand, do not have to be repeated at every individual set-up of the instrument, but, in a well-made instrument, can be relied on to retain their settings over long periods.

From the above the student will gather that the question of adjustment of his instruments is one of very great importance, and he will be well advised to make every endeavour not only to learn how to use his instruments and to care for them, but also to know the various adjustments necessary, and how to make them. In addition, he should note carefully the cases where the effects of errors of adjustment may be eliminated by suitable methods of observing. Every time he takes over a new theodolite or level, which are the surveyor's principal instruments, his first step should be to study how the instrument fits into its box or case so that he can replace it easily, and then, after a general examination and cleaning, he should take it out to the field to test its adjustments.

CHAPTER II

INSTRUMENTS FOR RANGING AND MEASURING STRAIGHT LINES

1. Operations involving Ranging.

One of the most common and essential operations in simple surveying is the ranging out of long straight lines on the ground. Briefly, this means putting in a series of pegs or other marks such that they all lie on a dead straight line joining two fixed points, or on a straight line running in some fixed direction starting from one fixed point. Another very common operation is setting out a line at right angles to another line. When great accuracy is needed, it is advisable to use a theodolite for carrying out these operations, but there are a great number of cases where the accuracy required does not demand or justify the use of very elaborate methods or instruments. Hence, before considering the equipment used for direct linear measurements, we shall describe some simple equipment and instruments commonly employed in connexion with the setting out of straight lines and of lines at right angles to other lines.

2. Pegs and Ground Marks.

Pegs are used to mark definite points on the ground either temporarily or semi-permanently. They may consist of prepared pieces of wood, of from 1 in. to 3 in. square and from 6 in. to about 3 ft. long, flat at one end and pointed at the other, so that they can be driven into the ground easily. The size depends on the use to which the pegs are to be put and the nature of the ground into which they are to be driven. If they are used in the tropics, and it is intended that they should remain in position for any length of time, it is best for them to be made from hard wood and to be creosoted. Even at home it is well to use creosoted pegs if they are intended to last for more than a week or two.

In bush or forest country where growing timber is plentiful and there are no restrictions on cutting it, pegs of nearly round section can be cut from standing trees and then pointed at one end and flattened at the other.

4

Iron pegs, made of cut pieces of iron rod or tube, about $\frac{1}{2}$ in. to $\frac{3}{4}$ in. diameter, or long wire nails, are sometimes used instead of wooden pegs. Iron pegs naturally last longer than wooden ones, but are more expensive and are troublesome to carry about.

Small concrete pillars are generally used when a station or point has to be permanently marked. These may be anything from 6 in. to 12 in. square and from 3 in. to 24 in. high, and are generally built *in situ* on some sort of concrete foundation.

It is often necessary to indicate on a peg or pillar the exact point to or from which measurements are to be taken, or over which an instrument has to be set. This is usually done by means of a nail, tack or brass stud driven into the flat top of the peg or let into the top of the concrete pillar.

3. Ranging Poles.

Ranging poles are thin poles used for marking a point in such a way that the position of the point can be clearly and exactly seen from some distance away. They consist of straight wooden or tubular metal rods of circular section, from 6 ft. to 10 ft. long, and tapering from a diameter of about $\frac{3}{4}$ in. to 1 in. at the bottom to about $\frac{1}{2}$ in. to $\frac{3}{4}$ in. at the top. At the bottom they are fitted with a heavy metal pointed shoe which enables them to be stuck easily in the ground. Usually they are painted in alternate bands of red and white, each band being a foot or a link deep, so that on occasion the pole can be used for the rough measurement of short lengths. When it is necessary to sight them from a long distance, they are often provided with two small flags of different colour, usually one red and the other white, fastened one above another near the top.

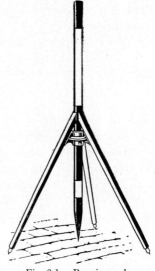

Fig. 2.1.—Ranging pole and stand

When a ranging pole has to be centred over a mark in a peg, and has to be left in position for some time, it can be guyed with wire or string stays fastened to pegs. Makers, however, can supply small folding tripods specially designed for supporting ranging poles over a mark. Fig. 2.1 shows a ranging pole supported by such a stand. The pole can be plumbed—that is, made vertical—by standing a few yards away from it and suspending a plumb bob about an arm's length in front of the eye with the supporting string in line with the ranging pole. The verticality of the pole can then be judged by seeing if its centre line and the line of the string of the plumb bob appear to be in the same vertical line. The test should be made from at least two directions at right angles to each other.

If, as is often the case, a ranging pole has to be held over a mark for a very short time only, as when a single observation is being taken to it, it can be held very approximately vertical by supporting it very loosely between the forefinger and thumb of the right hand, the hand being held slightly above the level of the eyes, so that the pole can swing comparatively freely about the points of support on the fingers as pivots, and can thus assume of its own accord a vertical position under its own weight. For this, it should be held with the point of the shoe not actually on, but very slightly above, the ground mark, the hand and rod being moved as a whole until the point of the shoe appears to be directly over the mark.

The chief defects from which ranging rods are liable to suffer are warping and becoming loose in the metal shoe. Warping can be detected by eye by sighting lengthways along the pole held at eye level, or the pole can be tested against a flat surface. Warped poles can be used for some rough purposes but they should never be used for work where accuracy is essential. Loose shoes can sometimes be made good by replacing the screws or nails fastening the metal of the shoe to the wood, or by hammering small wooden or metal wedges between the shoe and the side of the pole.

Ranging poles are occasionally provided with two short, narrow, vertical sighting slots passing through the centre of the section at right angles to one another and set about eye level. These are intended for roughly setting out lines at right angles to the main line, and are really a simplified form of cross staff (p. 9).

4. Plumb Bobs.

Plumb bobs are used for many purposes in surveying—for testing the verticality of ranging poles or signals; step chaining; transferring to the ground the end marks of steel bands; &c.—but one of their chief uses is to facilitate the accurate centring of a theodolite over a ground mark. The form of plumb bob usually employed consists of a short cylinder of brass, about 4 to 8 oz. in weight and 2 in. in length, and pointed at the lower end; or else a specially shaped piece of brass, shaped something like a pointed pear, round on top and pointed at the bottom. A small recess is bored out at the top to take the knot of a string, and a screw, with a hole bored through its centre vertically down its length, screws into this recess. The supporting string, which is usually about 5 ft. long, goes through the hole in this screw, and a knot, fitting into the recess below the bottom of the screw, prevents the bob from working off the string. A loop and sliding button at the other end of the string enables the effective length of the latter to be shortened or lengthened, and the plumb bob thus to be set and held at any desired height.

Some makers provide, at a slight extra cost, an extremely convenient form of adjustable plumb bob. In this, pressure on a small catch at the side of the bob itself releases the string and enables it to be wound or unwound about a roller inside the plumb bob, release of the pressure preventing further movement of the string. Hence, by this means, the length of the string in use at any time is easily controlled.

The point of a plumb bob should lie on a prolongation of the line of the string when the plumb bob is freely suspended. By hanging up a plumb bob to a suitable support and viewing it from a short distance away so that the eye can follow the line of the string to the point of the bob, it can easily be seen if this is so or not.

In mining and tunnelling work a couple of plumb bobs are sometimes suspended down a deep shaft to enable a line on the surface to be transferred to the underground workings. This means that the supporting strings must be very long, and in such a case the bobs are inclined to oscillate or to take a very long time to come to rest. Such oscillations can be damped down by letting the bob hang in a vessel containing oil.

In certain modern theodolites and other equipment the ordinary plumb bob is not, or need not be, used, plumbing over a ground mark or under a mark in the roof of a tunnel being done by optical means with a special attachment or instrument known as an " optical plummet ".

5. Line Ranger.

It is sometimes necessary to choose an intermediate point such that it lies on the straight line joining two distant points. If either of the distant points can be visited easily, and the line is not too long, an intermediate point can be lined in by eye by standing a short distance behind the point visited and signalling to a labourer to move

a ranging pole until it appears to the eye to be in line with the two
extreme points of the line.   Often, however, it is not convenient or
possible to visit either end of the line.   In this case an intermediate
point can be selected speedily and conveniently by means of a line
ranger.   This is a small instrument consisting of a couple of reflecting
surfaces arranged one above the other with their surfaces perpendicular to each other and their planes perpendicular to the base.   The
principle is very simple and will be understood from fig. 2.2.   The
ray from A is reflected by the surface **ab** in the direction oC so that
oC is at right angles to Ao, and the ray from B is reflected by the

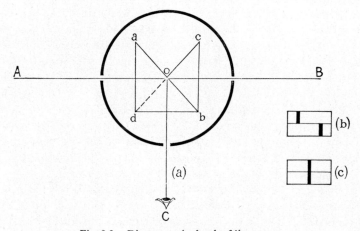

Fig. 2.2.—Diagrammatic sketch of line ranger

surface **cd** so that the reflected ray is at right angles to oB.   If o is
on the line AoB, the image of one point will appear to an eye placed
at C to lie above the image of the other and in a straight vertical line
with it as shown in fig. 2.2c.   If the point o is not on the line AB,
the two images will appear to be separated as shown in fig. 2.2b.
Hence, the instrument is used by moving it backwards and forwards
along a line approximately at right angles to AB until the images of
both points appear to touch and coincide.

One of the mirrors in the ordinary line ranger is adjustable, and
the adjustment is carried out by setting out three points on line,
holding the line ranger over the middle point and then seeing if the
images of the marks at the extreme points coincide.   If they do not
do so they are brought into coincidence by means of the adjusting
screw operating the adjustable mirror.

6. The Cross Staff.

It is very often necessary to lay off lines at right angles to another line. If great accuracy is desired, the work should be carried out by theodolite, but in many cases, notably when taking offsets from a chain line, something much more approximate, and easier and quicker in execution, is needed. For this purpose two small instruments are often used—the cross staff and the optical square.

The cross staff in its simplest form consists of a piece of wood or other material shaped as a cross (fig. 2.3). At or near the extreme

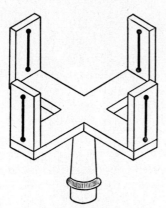

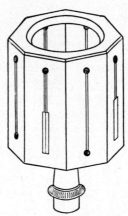

Fig. 2.3.—Simple cross staff

Fig. 2.4.—Ordinary octagonal cross staff

ends of one arm of the cross are fixed two sighting vanes made of metal, wood or vulcanite. Each of these pieces is provided with a narrow vertical slit in the centre of which is stretched a fine vertical wire. These two wires give a definite sight line to an eye placed near one slit. A similar pair of sighting vanes fitted at the ends of the other arm of the instrument gives a line of sight at right angles to the first. The instrument is supported on a very light collapsible tripod or on a pointed pole which can, if necessary, be stuck in the earth. Hence, the cross staff is set up at the point on the line from which the right angle is to run, and is then turned until one line of sight intersects or views a ranging pole or other mark on the main line. When the instrument is thus aligned, the line of sight given by the wires and slits in the other two vanes will be a line at right angles to the first line.

Another form of the instrument consists of a hollow octagonal box, about 6 in. deep and about 4 in. wide between opposite faces. Vertical sighting slits are cut in the middle of each face, such that the lines between the centres of opposite slits make angles of 45° with each other. It is therefore possible to set out angles of either 45° or 90° with this instrument. The principles involved, and the method of using this form of cross staff, will easily be understood from the description of the simpler form and from fig. 2.4.

A cross staff can easily be tested by setting it up at a point on a straight line from which a right angle has been set out by theodolite or other means, and then seeing if the angle set out with the cross staff agrees with the right angle already properly determined.

7. The Optical Square.

The optical square, being a small pocket instrument slightly larger than a watch, is somewhat more convenient than the cross staff for setting out a line at right angles to another, the object for which it has been designed. It consists in principle of a small round shallow box, with three narrow openings in the side at H, I and J (fig. 2.5). In line with the openings H and I is a piece of flat glass, **bb'**, unsilvered at the bottom and silvered on top with the silvered surface facing the opening H, this glass making any convenient angle with the line HI. Above the line HI is a mirror D set opposite the opening J and making an angle of 45° with the half-silvered mirror **bb'**. Now consider a ray of light FJE striking the mirror D at an angle ϕ. By the laws of reflection of light the reflected ray DB will make angle BDG $= \phi$ with the mirror D.

But

$$\text{angle DGB} = 45°$$

and so

$$\text{angle DBG} = 180° - (\phi + 45°)$$
$$= 135° - \phi.$$

Again, by the laws of reflection,

$$\text{DBG} = \text{HB}b,$$

so that

$$\text{HB}b = 135° - \phi,$$

and hence

$$\text{DBH} = 180° - 2(135° - \phi)$$
$$= 2\phi - 90°.$$

But
$$EDB = 180° - 2\phi.$$

Hence, in triangle DEB,

$$\text{angle DEB} = 180° - (EDB + DBE)$$
$$= 180° - (180° - 2\phi + 2\phi - 90°)$$
$$= 90°,$$

and it follows that the line FJE is perpendicular to the line HI. Consequently, if E and C are points on the straight line from which the

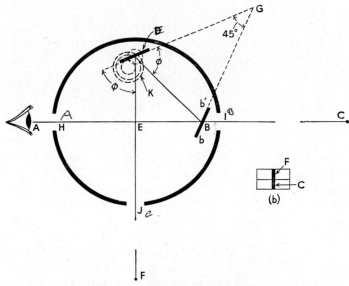

Fig. 2.5.—Diagrammatic sketch of optical square

right angle is to be set out, the instrument being held with its centre over the point E, and if the point C is sighted through the unsilvered part of the mirror **bb′** and the openings H and I, the image of any point F on the line EF at right angles to HC will appear in the instrument to lie directly above the point C. The instrument is therefore used by holding it over E to sight on a ranging pole held at C and then getting an assistant to move a ranging pole F about until the image of F, as seen in the upper silvered part of the mirror **bb′**, appears to be directly over a part of the ranging pole at C, as seen directly through the openings H and I and the unsilvered part of the mirror **bb′**. Fig. 2.5*b* shows the appearance in **bb′** when a portion of a ranging pole and its reflected image appear to coincide.

The optical square can easily be tested and, if necessary, adjusted by the following method. Standing at any intermediate point O on the line AB, sight a pole held at the point A and get an assistant to put in a peg or mark, **a**, such that this mark appears in the instrument to be at right angles to OA (fig. 2.6). Now turn round to face

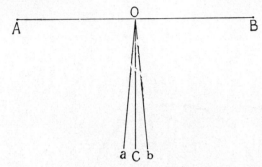

Fig. 2.6.—Adjustment of optical square

point B and turn the instrument over in the hand. Get an assistant to hold a ranging pole over the point **a** sight a pole held at B, and, if the instrument is in adjustment, the image of the pole at **a** will coincide with the pole seen at B. If the instrument is not in adjustment, it will be necessary to move the ranging pole at **a** sideways to make its image coincide with the image of the pole at B. Let **b** be the new

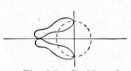

Fig. 2.7.—Position of observer when using optical square.

position of the ranging pole. Bisect the distance **ab** at the point C. Then OC is the true perpendicular at O to the line AB. The instrument can now be adjusted by means of a small milled head (K in fig. 2.5), which is usually fitted in the lid and can be used to rotate the mirror D until B and the image of C appear to come together.

Since a stand or tripod is normally not provided with an optical square, care has to be taken to hold the instrument over the exact point on the chainage line from which the right angle is to be laid out. The best plan is to stand facing one end of the line with the heels touching and the feet making an angle of about 45° with one another, and with the middle point of the line joining the toes coinciding with the point at which the right angle is to be laid out (fig. 2.7). By bending the head forward a little, the instrument, when held to one eye, will be approximately over the proper point.

8. The Prism Square.

The prism square is a simplified form of optical square and consists of a prism of glass, ABCDE (fig. 2.8), such that the angle between the faces AB and AE is 90°, and that between the faces BC and ED is 45°. The object P is seen directly through a peep hole on top of the prism, while the object R, on a line at right angles to the line QP, is seen underneath P after reflection from the faces ED and BC.

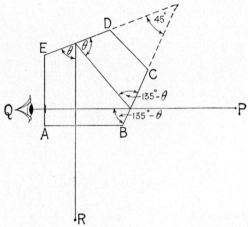

Fig. 2.8.—Diagrammatic sketch of prism square

There are no means of adjusting this instrument and the surveyor is dependent on the maker for providing a prism with angles of 90° and 45° between the sides concerned.

9. The Linen Tape.

The linen tape used by surveyors is the ordinary linen tape which is in common use by builders, architects, plumbers, &c., and hence should be familiar to the reader. It consists of a long strip of best-quality yarn, about $\frac{5}{8}$ in. wide, which can be wound into, or unwound out of, a flat circular case of leather, metal or bakelite. One end of the strip is attached to a small metal cylinder or drum working inside the case, and a small metal loop is fastened to the other end outside the case, the tape itself passing through a narrow opening in the side of the latter. Each end of the drum revolves in a metal bearing, and one end passes through the wall of the case to be attached on the outside to a handle which, when the tape is not in use, folds down on top of the case. The handle is used to turn the drum and hence to wind or unwind the tape. The latter is usually 50 ft., 66 ft. or 100 ft. long, and is graduated in suitable units of length, often with feet, inches and half-inches on one side and links and poles on the other.

The linen tape is used in surveying mainly for the measurement of short distances, such as offsets in chain surveying, where great accuracy is not needed. It has the disadvantages, however, of fraying very easily, particularly at the end to which the loop is attached, and of altering its length very considerably when it gets damp. For these reasons, tapes can be obtained in which strands of copper wire are woven into the linen, and such tapes are generally called *metallic tapes*. These tapes last longer than the ordinary linen kind and do not alter their lengths quite so easily, but, like linen tapes, they should not be used for very accurate measurements. Refill linen and metallic tapes can be obtained and are easily fitted into the cases to replace torn or damaged tapes. In addition, repair lengths with metal loops are obtainable which can be sewn to a torn tape at the point where the break occurs.

THE SURVEYOR'S CHAIN

10. General Description.

The two instruments most commonly used for the direct measurement of the lengths of survey lines are the surveyor's chain and the long steel band. The chain is not capable of giving such accurate results as the steel band and cannot be used for really refined work, but it is much less easily damaged, is easier to repair, and requires less care than the steel band both in handling and in maintenance. For some kinds of very rough work such as route or exploratory surveys, topographical surveys for mapping on very small scales, &c., distances are sometimes measured with wire or hempen ropes, pedometer or " perambulator ". These, of course, give nothing like the same degree of accuracy as the chain or steel band.

Surveyor's chains are usually either 66 ft. or 100 ft. long, a chain of 66 ft. often being called a *Gunter's chain* or *link chain*. The length of a 66-ft. chain is a definite unit of length, the *chain*, each of the 100 subdivisions being known as a *link*. When a survey is made with the specific object of determining areas of land, and these areas are to be defined in terms of acres or its subdivisions, it is usually more convenient to use the link chain rather than a chain divided in feet because of the simple relation 10 square chains equal 1 acre. There is no such simple relation between square feet, or their multiples, and the acre. In addition, there is the simple relation between the chain and the mile, as 10 chains are equal to 1 furlong, or 80 chains to the mile. Link chains are therefore generally favoured by land surveyors, but on most modern engineering work it is customary to use a chain divided in feet. A chain much longer than 100 ft. would be difficult to handle in the field.

The ordinary chain, whether 66 ft. or 100 ft. long, consists of 98 pieces of No. 8 to No. 12 S.W.G. wrought-iron wire formed into the shape of long links, with rings at their ends so that they can be joined together by short links of the same material. When the 98 pieces

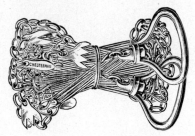

Fig. 2.9a.—Folded up chain
(By courtesy of Messrs James Chesterman & Co., Ltd.)

are connected together, the two links at the ends are joined to shorter links which, in turn, are connected to brass handles (fig. 2.9). A small brass *tally* attached at the end of every tenth link divides the chain into ten equal parts, and enables the reading on the chain to be obtained without any difficulty.

Fig. 2.10a (p. 17) shows the end links of a chain with the attached handle, and fig. 2.10b shows the method of attaching the tallies at every tenth link. The general shapes of the different tallies are shown in fig. 2.10c. These tallies are made of small pieces of brass, and those at the 10, 20, 30 and 40 marks are pointed so as to have different numbers of teeth, one tooth denoting 10, two teeth 20, three teeth 30, and four teeth 40. The 50 mark is round-shaped, and the 60, 70, 80 and 90 tallies are shaped exactly the same as the 40, 30, 20 and 10 tallies respectively.

When not in use, the chain is folded up from the middle in such a way that it forms a small bundle, shaped something like a sheaf of wheat or double cone, with the handles outside. A strap, fastened round the waist, keeps the whole chain together in its folds (fig. 2.9a).

The overall length of a chain is generally to be taken as the distance between the outsides of the handles, but sometimes it is the distance measured between the insides. Fig. 2.10a illustrates the latter case and fig. 2.9b the former. In either case, a measurement of the end link or foot lengths with an ordinary footrule will indicate which way to take the measurement. Occasionally there is a small semicircular recess, approximately the same diameter as an arrow, in the

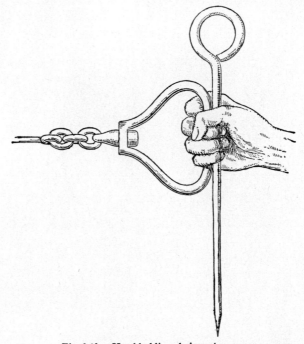

Fig. 2.9b.—Hand holding chain and arrow

centre of the outside of the handle, and, if so, the chain should be held with the arrow fitting into this recess. The other points of division are taken as the centre of the middle of the small links connecting the longer links. In reading a chain, single links are counted forwards to, or backwards from, the nearest tally mark to give the nearest single link or foot; tenths of a link or of a foot are estimated by eye.

After they have been in use for some time, chains are liable to appreciable alterations in length. Kinks or slight bends in the longer links will cause shortening, and opening out of the shorter links, as well as stretching of the wire of which the chain is made, will cause lengthening. Hence a chain should be tested at fairy frequent intervals

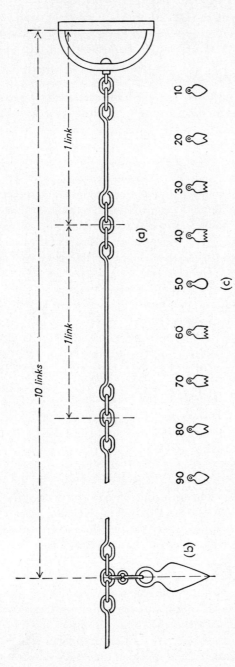

Fig. 2.10.—(*a*) End of chain with handle; (*b*) Method of attaching tallies to chain; (*c*) Tallies used on chain.

on a standard of length laid out with a good steel band. A chain that is too long will make the measured length of a line too short, and a chain that is too short will make the measured length too long.

11. Adjustment and Care of Chain.

Chains can be adjusted to their proper length by means of the short links connecting the longer ones. Before doing anything else, the surveyor should examine the chain from end to end and straighten any bent links and close any that are open. Then, if the chain is still too long or too short, one or more short links can be removed or others added to make the over-all length correct.

Although the chain will stand up to much rougher usage than the steel band, it should not be used carelessly. When it is pulled along the ground there is a tendency for the links to catch on twigs or stones. Consequently, the pull on it should be eased the moment any resistance is felt and the caught-up link released. After use, it should be dried if it is wet, and all adhering clay and dirt removed. It should not be bundled up carelessly but should be " folded up " carefully as described below, and the strap holding it together firmly fastened. If it is not going to be used for some time, it is advisable to cover all the links with Vaseline before folding it up and putting it away; then, when it is to be used again, the Vaseline should be removed after the chain has been extended to its full length.

12. Unrolling and Rolling-up the Chain.

To unroll the chain, remove the strap and take the two handles in the left hand, holding the remainder of the chain (with the exception of the last two or three links at each end) with the " waist " firmly grasped in the right hand. Still holding the handles with the left hand, throw the remainder of the chain forward by means of a sharp underhand sweep of the vertically extended right arm. Lay the handles on the ground so that the two halves of the chain lie side by side with both handles close to the feet. Leaving one handle where it lies, pick up the other handle and walk forward until the chain is fully extended.

To roll up the chain, lay the two handles together on the ground so that the chain lies doubled up with the forward and rear lengths lying side by side, the 50-tally being at one end and the two handles at the other. Pick up the two links on either side of the 50-tally—that is, the 50th and 51st links—and hold them side by side in one hand. With the other hand grasp the next two links—the 49th and

52nd links—and, holding them together, lay them at a slight angle across the first two. Similarly, take the next pair of links and lay them together at a slight angle to the second pair. Proceed in this way, placing each successive pairs of links at a slight angle (with the angles all in the same direction) to the previous pair so that, as the links are folded up, the chain takes up the usual shape of a double cone with a waist in the middle. Fasten the strap firmly around the waist when the last two links carrying the handles are in position.

13. Using the Chain.

At least two men are necessary to use the chain, one to each handle. The man at the forward end is known as the front or forward chainman, or leader, and the man at the rear as the rear chainman, or follower. The latter, who is often the surveyor himself, is responsible for seeing that the front chainman puts in the arrows marking the end of every chain length on the proper alignment. The rear chainman also is usually responsible for keeping the field book unless this is kept by the surveyor or by a special booker.

Each of the arrows used to mark the chain lengths consists of a piece of No. 6 to No. 8 S.W.G. wrought-iron wire, about 12 in. long, one end being pointed and the other shaped into a circular ring about $1\frac{1}{4}$ in. diameter. A small piece of coloured rag is usually fitted to the ring so as to make the arrow easily visible.

When starting the chaining of a line, the leader should have ten arrows and a ranging pole in his possession, and the follower, if his end of the line is not already marked by an arrow, should have one arrow in addition to a ranging pole. The follower allows the leader to go forward until the chain is fully extended, when he signals him to stop; then, by signalling with his hand, he gives him the approximate direction of the line, the leader, if necessary, using his ranging pole to obtain the correct alignment. Both men then hold their respective handles and shake the chain gently up and down until it lies straight and free from all obstructions. The leader, holding an arrow vertically against the handle of the chain as shown in fig. 2.9b, and standing a little to the side of the line, bends down and looks towards the follower, at the same time holding the arrow and handle at ground level but clear of his body, so that the follower can see past him to the terminal point of the line.

Meantime, the follower holds his handle against the mark or arrow at his end, and, sighting the end point of the line, signals to the leader to move his arrow to one side or the other until it is dead on line. When he receives the signal that his arrow is on line and observes that the chain is straight, the leader, holding his end of the chain taut, sticks his arrow in the ground, picks up his handle and the remaining arrows and proceeds to draw the chain forward. As the chain moves forward, the follower picks up the arrow at his end and moves forward to the arrow left in the ground by the leader, and, just before he reaches it, shouts to the leader to stop. The operation is now repeated for each chain length until the whole line has been measured.

When ten chain lengths have been laid out, the leader will have no arrows left in his hand. He therefore waits for the follower to come up and hand over to him the ten arrows that should now be in the follower's possession, the number of arrows so handed over being carefully counted by each man. This counting of the arrows at every tenth chain length is very important, as it helps to prevent the occurrence of gross errors due to dropped chain lengths and arrows being left in the ground and lost.

When the end of the line is reached, the end point will not normally be exactly at the end of a chain length, and it will be necessary to measure an odd length consisting of a part only of the full length of the chain. If there are no obstacles to prevent this, the chain should be dragged past the end mark until it is fully extended with the rear handle against the last arrow. The surveyor can then walk along the line, and, by means of the tallies, can note and book the exact chainage of the end point of the line.

14. Chaining on Sloping Ground.

When the plan is plotted it should represent the features on the ground as projected on a perfectly flat horizontal surface. Accordingly, the distances required for plotting purposes should be the horizontal projections of all sloping lines. There are two ways of obtaining these. One is to make all measurements in such a way that they represent true horizontal distances. The second is to measure the angle of slope, and then to apply computed corrections to the measured distances to reduce them to the equivalent horizontal distances. The first method is that usually employed when working with the chain. The same method is also sometimes used when working with a long steel band, and great accuracy is not needed, but, as the steel band is usually used when accuracy greater than that possible with the chain is desired, it is more usual, when working with it, to measure the slope and apply a computed correction for slope later.

The method of chaining " on the horizontal "—known as *drop-* or *step-chaining*—will be understood from fig. 2.11. One end of the chain, or some convenient subdivision, is held at ground level at **a**, and a subdivision or end is held at **b** in such a way that the chain appears to the eye to be horizontal. The point **b** is then transferred to the ground at **a′**, vertically below **b**, either by means of a ranging pole held vertically as nearly as can be judged by eye, or, better still, by using a plumb bob, an arrow being placed at **a′**. Similarly, part of the chain being held on the ground against the arrow at **a′**, the point **a″** is laid out so that **a′b′** is the horizontal projection of **a′a″**. Again, the point **a‴** is marked so that **a″b″** is the horizontal projection of **a″a‴**. The length of the line **aa‴** is then taken as **ab** + **a′b′** + **a″b″**.

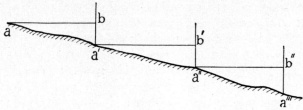

Fig. 2.11.—Chaining on sloping ground

The above description applies to a line chained down-hill, in which case it is the leader who holds the chain above the ground, but the procedure in chaining up-hill, in which case it is the follower who holds the chain above the ground, is very similar and will easily be understood.

As an alternative to using a ranging pole or plumb bob for transferring vertically the point on the chain to the ground, a special *drop arrow* can be obtained from instrument makers. This consists of an ordinary arrow with a ball of heavy metal fitted near the pointed end. Hence, if this arrow is dropped from a point on the horizontally held chain, it falls and sticks in the ground vertically below the point from which it is dropped.

15. Accuracy of Chain Measurements.

The accuracy of chaining is expressed as the ratio of the error at the end of a measured line to the length of that line. There are a number of factors which control the accuracy of chaining, but the main ones depend on the state of the chain, its true length as compared with its nominal length, the slope of the ground and the roughness or smoothness of its surface, and the manner in which the chain is used, particularly the method of dealing with sloping ground. At one end of the scale the accuracy may not be more than 1/100, or one part in a hundred, but good chaining with a standardized and adjusted chain in good order may be expected to give an accuracy of somewhere between 1/500 and 1/1000.

THE STEEL BAND

16. General Description.

The steel band consists of a long narrow strip of blue steel, of uniform width and thickness, and suitably marked or graduated for measuring the lengths of long survey lines. Steel bands may be anything from 50 to 500 ft. long, $\frac{1}{16}$ in. to $\frac{3}{4}$ in. wide, and from 0·01 to 0·03 in. thick, and they may be graduated in feet, links or metres. Probably the length most commonly used in the Empire and at home is 100 ft., but bands 300 ft. long are also extensively employed.

As a general rule, steel bands are provided with handles at each end, similar to the handles used on ordinary chains. These are fastened by a metal ring to a small metal loop or closed clip riveted to the end of the band. Sometimes, however, handles are not used, their place being taken by a leather thong or a piece of strong string.

In some bands, the nominal length is measured between the ends of the handles, just as the length of an ordinary chain is often measured. In other cases, the zero and end graduations are on the tape itself and the handles are thus not included in the measurements but are used simply for handling and stretching the band. Intermediate graduations are either etched directly on the surface of the band or else they are marked by small brass studs or sleeves passing through or fastened to the metal. As a general rule, the wider bands have the graduations etched on the surface and the narrow ones the subdivisions marked by studs or sleeves.

The manner of graduation or subdivision also shows considerable variations. Bands of 100 units (feet, links or metres) or less are generally marked at every unit, and, if studs are used, there is a flat numbered sleeve at the tens, with a stud larger than the others at the odd multiples of five. In such a case, the first ten units are generally subdivided into tenths, the tenths being marked by very small studs. With longer bands of 200, 300 or 500 units, there is generally a numbered mark at the end of every 100 units, the first hundred units are divided into tens, the first ten into single units, and the first unit is subdivided into tenths.　Occasionally the band is extended for one unit length to the negative side of the zero mark, and this unit is divided, like the unit to the positive side of the zero, into tenths.　Etched tapes sometimes have the first unit graduated to hundredths, but, where the graduations are marked by studs, the hundredths of a unit are obtained by estimation.

For convenience in handling and carrying, steel bands are almost invariably wound on special steel crosses or metal reels from which

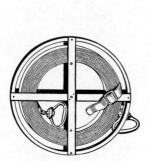

Fig. 2.12a.—Steel band
on cross
(By courtesy of Messrs James
Chesterman & Co., Ltd.)

Fig. 2.12b.—Steel band on reel
(By courtesy of Messrs. James
Chesterman & Co., Ltd.)

they can easily be unrolled.　Fig. 2.12a shows a band on a steel cross and fig. 2.12b a metal reel.　An arrow passed through a hole in the centre of the cross serves as an axis about which the cross can be turned and the band wound on or unwound.　In the case of the reel, the hand is passed through the strap shown in the figure, and the reel can be turned about its axis by means of a knob attached to one of the flanges.　Means are provided for passing one of the handles inside the reel and attaching it to an internal revolving drum.　Consequently, the band will roll on the reel or be unrolled from it as the flange with the knob is turned.　When the tape is fully wound up, the end is fastened to the cross or reel by means of a strap passing through the handle.

The choice of the most suitable length, width and thickness of a steel

2

band is decided largely by the way in which it is to be used. Of the two methods of use the most common consists of straightforward measurements along the surface of the ground, exactly like ordinary chaining. The second method, which is normally the more accurate of the two, consists of measurements *in catenary*. In this method, the band is suspended above the ground and allowed to hang in a natural curve (the catenary) under its own weight. As a general rule, bands longer than about 100 ft. are not used for surface measurements, and for such work, whatever length is chosen, the band should be of good strong section, say $\frac{1}{2}$ in. wide by 0·03 in. thick. For catenary measurements, bands up to 300 ft. or so are employed, and these should be of light section, say $\frac{1}{8}$ in. wide and 0·01 in. thick.

17. Adjustment and Care of Steel Bands.

If a band is provided with handles, and the nominal length is to be taken between the ends of the handles, some slight adjustment is sometimes possible by closing up or opening the links which attach the handles to the band. Ordinarily, however, the over-all length of the band is between marks actually on the metal, and in that case, unless the old marks can be conveniently obliterated and new ones made, the best that can be done is to find the true length between the marks and apply a correction to the field measurements to reduce them to the values they would have if they were made with a band of correct length. The true length is, of course, found by comparison with a standard of length or with a band already standardized.

Steel bands, as has already been noted, are more easily damaged or broken than the ordinary surveyor's chain and therefore require greater care in handling and storage. If they are used for surface measurements, particular care should be taken in dragging them along the ground and sudden jerks and pulls should be avoided. Before use, they should be unrolled, cleaned, and examined for kinks, and, if any are found, they should be straightened out as far as possible. After use, bands should again be examined for kinks and then carefully cleaned and dried, and, if they are not to be used for some time, they should be well and carefully greased with non-corrosive Vaseline before being wound up on their reel.

18. Repair of Steel Bands.

Makers supply different types of outfits for repairing steel bands. One consists of a special punch and a supply of rivets. A scrap piece of band is used, and this is riveted over the two broken ends. In

another, a special metal sleeve coated on the inside with soft solder
and flux is provided. The broken ends of the band are brought together
inside the sleeve, and a lighted match applied underneath the latter
so that the solder melts and joins the sleeve to the ends of the band,
both of which must be well cleaned beforehand. A third consists
of a small metal sleeve with small screws at each end and a hole,
through which the contact of the broken ends can be seen, in the
centre. The ends of the band are brought together in the sleeve, and
the pressure of the screws, when these are tightened, prevents the
two pieces from coming apart and holds them firmly in position.

19. Unrolling and Rolling the Band.

When unrolling the band, one chainman takes the free end of the
tape and walks slowly forward, while the other chainman holds the
arrow on which the cross is to revolve, or the strap of the reel, so
that the band comes off freely and easily and without jerks. The man
at the cross or reel end also sees that the band does not come off too
quickly, and that it does not get caught up on obstacles. As his end
of the band comes in sight, he warns the man at the other end to be
prepared to stop, at the same time preparing to give way a little at
his end in case of a sudden jerk. When the end of the band is reached,
he gives the signal to stop, and both men then lay their ends on the
ground ready to start work.

In rolling up, the band is first laid out straight on the ground
and one end is fastened to the cross or reel. The man working the
latter now starts to walk slowly towards the other end of the band,
at the same time winding it up as he goes along. In doing this, he
sees that it is not caught on any obstruction and is wound firmly and
tightly without any sudden jerks taking place. When the end is
reached, it is fastened to the cross or reel by means of a strap or
string.

20. Surface Chaining with the Steel Band.

In its simplest and most fundamental form, the procedure in
surface chaining with the steel band is almost exactly similar to that
used with the ordinary surveyor's chain, although the accuracy obtainable is generally a little higher. For many engineering surveys the
chaining is often done in this way, step-chaining being used as before
over sloping ground, and no attempt made to maintain a constant
known tension. All steel bands, however, are only of correct length
when they are at a certain tension and temperature—usually 10 to

School of
Engineering
Limerick Technical College

15 lb. and 60° F. respectively—and the tension and temperature at which they are of standard length are often stamped by the makers on the handle or on a metal tag attached to the handle. Accuracy may therefore be increased if the band is used at the tension at which it is supposed to be of standard length, if temperatures are measured and recorded, and computed corrections applied to the field measurements to allow for temperature effects, and if, instead of employing step-chaining to counteract the effects of slope, the measurements are made along the sloping surface, the angle of slope is observed, and a correction applied to the measured length to reduce it to the equivalent horizontal length.

A constant and known tension can be maintained if the pull at one end is taken through a spring balance. Spring balances for this

Fig. 2.13.—Spring balance
(By courtesy of James Chesterman & Co., Ltd.)

purpose can be obtained from instrument makers. They are usually of the barrel type, registering pulls up to about 15 to 25 lb., with a special clip at one end to fasten to the handle or end of the band, and a handle at the other end (fig. 2.13). Hence, the chainman, instead of pulling direct on the handle of the band, hooks on the spring balance and pulls on the handle of the latter until the sliding index shows that the tension stands at the desired value. He holds this tension while measurements are in progress.

Temperatures are measured, usually in degrees Fahrenheit, on thermometers made for the purpose. These thermometers are enclosed in a metal tube to protect them against damage, one side of this tube having a long, straight, narrow opening cut in it through which the graduations on the thermometer stem can be seen and read. In ordinary surface taping, only one thermometer is used, and it is placed on the ground alongside the band a minute or two before the reading is taken. Sometimes, for work of the highest accuracy, readings are taken for every stretch of the band; but often, for ordinary work, the thermometer is only read once or twice during the course of the day, a mean temperature for the day being used where there are several readings, and all corrections worked out for this mean.

Angles of slope are usually observed by the rear chainman, or follower, by means of a clinometer or Abney level or similar instrument (see Chap. V, §§ 2–4), sights being taken to a ranging pole held by the front chainman, or leader. The follower notes, or makes, a particular mark on the ranging pole held by the leader on which to sight, and he supports the clinometer or hand level against a mark at the same height on his own ranging pole. Thus, the line of sight will be a line parallel to the sloping surface of the ground. Angles of slope should be observed at all points where there is a decided change in the slope of the ground, even if this occurs at some intermediate point of a chain length. The chainages of all points where slopes are observed must, of course, be noted in the field book.

The methods of calculating and applying corrections for tension, temperature, slope, &c., are described in *Principles of Surveying*, Chap. VI.

21. Accuracy of Surface Chaining with the Steel Band.

Provided the ground over which chaining takes place is reasonably smooth and clear of obstacles tending to make the band lie unevenly or out of its proper alignment, and that the band has been properly standardized, an accuracy of something like 1/1000 may be expected for simple ground chaining in which no spring balance is used, no temperatures are observed, and sloping ground is step-chained. If, on the other hand, the band is used with a spring balance to maintain the appropriate constant tension, and temperatures and slopes are carefully observed, an accuracy of anything between 1/1000 and 1/10,000 may normally be obtained without much difficulty. The highest degree of accuracy will only be reached on reasonably flat ground on which all undergrowth has been cleared to ground level, and in which the band has been used in such a way as to eliminate errors due to the thickness of the arrows, or this thickness has been allowed for when reducing the final results.

22. Catenary Chaining with the Steel Band.

This method, if carefully executed, will usually give more accurate results than surface chaining unless exceptional precautions are taken with the latter. Apart from the question of accuracy, however, catenary chaining is particularly suitable for working over rough country or cut lines through dense forest or heavy undergrowth, because in surface chaining it is necessary to clear everything down to ground level whereas in catenary chaining it is only necessary to clear down to a height of a foot or two above ground level. On long lines, this means a considerable saving in the amount of clearing to be done.

Bands of light section are used for catenary chaining, as heavy bands are difficult to handle with the tensions necessary to avoid deep sags at the centre of the span. A convenient section is $\frac{1}{8}$ in. by 0·01 in., used with a tension of 15 lb. to 20 lb. Lengths may vary from 100 ft. to 300 ft., the shorter lengths being the more convenient for straightforward chaining without a theodolite, and the longer lengths when measurements at one end are made to the axis of a theodolite in the manner described later. In all cases of catenary chaining, the bands used should be ones with the end marks on the bands themselves, not ones whose over-all length is measured between handles, and it will generally be more convenient if the handles are replaced by stout leather thongs.

The simplest method of using a band in catenary is for each chainman to support it about waist high, with both ends at the same height above the ground, the follower applying tension through a spring balance held in his right hand. With his left hand he holds a plumb line with which he makes certain that the mark on the tape is vertically above the mark on the ground. Meantime, the leader holds his end steady against the applied tension with one hand, and, when the follower signals that everything is ready, he transfers the forward mark on the band to the ground by means of a plumb bob held in his other hand.

In order to obtain the proper slope correction, it is necessary that both ends of the band should be held at the same height above the ground mark. This can be ensured if each chainman holds a ranging pole in the hand used to apply or hold the tension. The ranging pole will serve not only to indicate the height at which to hold the band, but also, if the shoe is stuck lightly in the ground, it can be used to act as a lever helping to steady the band.

An alternative method of working which avoids the use of plumb bobs is to have long, stout, wooden pegs driven ahead at every point where the end of a band will come. The tops of these pegs should be about $2\frac{1}{2}$ ft. to 3 ft. above ground level, and their centres should be marked by stout nails. The follower applies the tension through a spring balance and holds his end mark on the band against the centre of the nail in the rear peg. At the same time, the leader, while holding the tension, reads the small difference in length between the

end mark on the band and the centre of the nail. To enable him to
do this, either his end of the band can be suitably graduated on both
sides of the end mark or else the difference can be read by means of
a graduated foot-rule.

In measuring fractional or odd lengths at the end of a line, it will
generally be necessary to hold the band at some intermediate point.
This can be done by means of a tape grip, such as the Littlejohn Roller
Grip shown in fig. 2.14. The band is passed through a narrow slot
in the side of the grip so that it lies underneath the roller. When
tension is applied, the roller tends to wedge itself tightly between the
inclined top of the grip and the band, so holding the band firmly.

In both of the above methods, slope and temperature can be
measured in the ordinary way, care being taken in the organization

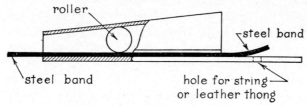

Fig. 2.14.—Littlejohn tape grip

of the work to see that the line of slope measured is the actual slope
of the line joining the end marks of the band.

The third method of catenary chaining involves the use of a
theodolite, and setting it up at the end of every alternate length of
the band. The latter is held with one end mark against or near a
centre mark on the horizontal axis of the theodolite, and with the
other end mark against a nail driven into a tall stake or peg about
3 ft. high, the theodolite being sighted so that the cross hair of the
telescope appears to intersect the nail. Tension is applied through
a spring balance at the theodolite end of the band, the handle of the
balance and the loop of the leather thong at the far end being held
against ranging poles used as levers to steady the band. The small
difference in length between the end mark of the band and the mark
on the theodolite axis is read either by a graduated foot-rule or else
direct from the graduations on the band, the mark at the other end
of the latter being held against the centre of the nail in the peg. With
lines of ordinary length, the error in taking the measurement to the
end of the horizontal axis of the theodolite instead of to the centre of
the latter is negligible.

It will readily be seen that time and labour will be saved if long bands are employed with this method. As a general rule, bands of 300 ft. are used, but with this length some form of intermediate support is necessary. Hence, the band is supported either at the 100-ft. and 200-ft. marks or else at the 150-ft. mark by means of forked sticks, which are held by labourers with the band supported freely in the forks. These sticks are carefully lined in both horizontally and vertically by sighting from the theodolite so that the points of support lie on the line joining the theodolite to the nail in the peg, the angle of slope being read on the theodolite.

It is an essential feature of this method of chaining for it to be possible to lay the mark on the band against a mark on one side of the horizontal axis of the theodolite. In older types of instrument this is nearly always possible, the small centre punch used by the manufacturer in turning up the axis on a lathe serving as a convenient mark. In some of the newer types of theodolite, however, micrometers, &c., prevent a band from being brought into contact with one end of the horizontal axis and the other end is covered in or is otherwise inaccessible.

This is probably the most accurate of all ordinary methods of chaining over rough ground not involving the use of special base-line measurement equipment. Further details will be found in Vol. I of Clark's *Plane and Geodetic Surveying for Engineers* (third and fourth editions). Although the frequent setting up of a theodolite does lead to some loss of time, with a well-trained party this is not excessive, and is justified by the added accuracy obtainable. It is, in fact, possible to complete two or three miles in a single day, provided the work is set out beforehand by putting in a peg or stake at every point where the end of the band will come.

23. Accuracy of Catenary Chaining.

The degree of accuracy to be expected from the first method of catenary chaining may be taken as anything between 1/1000 and 1/5000, and with the second method between 1/1000 and 1/10,000, according to the amount of care taken, the accuracy of the standardizations, and the way in which temperatures and slopes are measured, and the appropriate corrections applied. When the third method is used, the accuracy obtainable may be anything between 1/5000 and 1/50,000.

BASE MEASUREMENT APPARATUS

24. The Special Problems of Accurate Measurement.

The measurement of a geodetic base line involves the most precise measurement of the length of a line that a surveyor is ever called upon to undertake, and, although the subject of geodesy as a whole and the methods used in it are outside the scope of this book, the reader should have some idea of how a base line is measured, and of the equipment used.

The great difficulty in any very accurate determination of the length of a base line has always been connected with the establishment of the true temperature of the apparatus involved in the measurement and of the errors involved in the determination of the temperature. All ordinary materials expand or contract as they become hotter or colder. Consequently, a bar, chain or band used for base measurement will vary in length as its temperature changes. If temperature could be determined accurately, and the coefficient of thermal expansion of a bar or band were known, the effects of variations of temperature could be calculated exactly and allowed for. Unfortunately, this cannot be done, not easily at any rate, as there is no certainty that a bar or band stretched along the ground or suspended in the air under ordinary field conditions is at the same temperature as that registered by a thermometer placed near it, or even in contact with it or with the ground. The apparatus used, whatever its form, must have its true length determined at some known temperature, and it will no longer be of that length at a different temperature. Hence, if we do not know the exact temperature of the apparatus when it is used for measurement, we cannot apply the proper corrections to lengths measured with it. From this it follows that the best type of apparatus to employ is one in which variations in length, corresponding to variations in temperature, are very small.

25. Use of Invar for Bands and Wires.

In order to overcome these difficulties, many different types of apparatus—wooden rods, compensating bars, &c.—were used until fairly recently. All of this apparatus was exceedingly cumbersome and work with it was very slow. In 1896, however, Guillaume discovered a nickel-steel alloy, called *Invar*, which had a very low coefficient of expansion—seldom more than about one-tenth of that of steel, and often very much less. It was found possible to make long wires and bands out of this material, and the discovery revolutionized the measurement of geodetic base lines, with the result that the older forms of apparatus have become obsolete.*

While the chief advantage of Invar is its very low coefficient of thermal expansion, this coefficient is not the same for every Invar band, but varies a good deal with individual bands. In some it may even be negative, so that a band with a negative coefficient will actually decrease in length with an increase in temperature, instead of increasing. In general, however, the coefficient is positive, and seldom exceeds in value 0·0000005 per 1° F.—that of steel bands usually being about 0·0000062 per 1° F.—its mean value being somewhat less than this. The other great advantage of Invar is that bands and wires made of it enable base lines to be measured very much more rapidly and conveniently than they could be measured with any of the old types of apparatus, and at the same time just as accurately. Against this, however, Invar bands and wires have certain disadvantages. Besides being much more expensive, they are much softer and more easily deformed than steel bands and wires and hence require great care in handling. This applies particularly to wires. Moreover, Invar is subject to a phenomenon known as *creep*, which means that it undergoes a small, but fairly regular, increase in length as time goes on. This creep is removed to a considerable extent by a process of annealing known as *étuvage*, or *ageing*, but it never vanishes completely. In addition, it has been found that the coefficient of thermal expansion is also subject to a slow change with time. For these reasons, therefore, bands or wires made of Invar should have both their lengths and coefficients of expansion determined at intervals in a good standardizing laboratory, such as the National Physical Laboratory, particularly just before and just after an important base measurement.

26. Comparative Advantages of Bands and Wires.

When Invar was first employed in connexion with base measurements, it was used in the form of wires 1·65 mm. diameter and 24 m. long. The end marks of the wire were carried on small pieces of metal of triangular section, about 5 cm. long, which were fitted to the ends of the wire. These pieces of metal, called *réglettes*, were graduated in

---

* Descriptions of the older forms of base measurement apparatus, now mainly of historical interest, will be found in most standard textbooks on geodesy.

millimeters, and were used to measure the small distances between the end of the wire—taken as the zero of the scale on the réglette—and a mark carried on a *measuring head*. In practice, wires were very inconvenient and difficult to handle, as they tended to coil back on themselves and so to get damaged extremely easily. Hence, they have now given way to a considerable extent to long bands made of Invar, either 25 m. or 100 ft. long.

A band made of Invar is very similar in appearance to a band made of steel, except that Invar is of a light silver colour, whereas the ordinary steel band has a somewhat bluish tinge. As a general rule, also, the drums or reels on which Invar bands are wound have a much larger diameter than the diameter of the ordinary reel used for steel bands—generally three or four times larger. An Invar band is likely to be damaged if it is wound on too small a reel.

Invar bands, being much more easily damaged than steel bands, require much greater care in use and maintenance. All possible precautions should be taken at all times to guard against sudden jolts, and to prevent bending or kinking: for this reason, when they are being carried forward from one bay to the next, Invar bands should be supported at intervals, well clear of the ground, by slings or hooks carried by carefully drilled labourers. When not in use, they should be kept greased with non-corrosive Vaseline and well cleaned immediately before use. Leather thongs, and not string or cord, should be used at the ends or to fasten them to a reel, as a string or cord in contact with Invar is liable to set up corrosion.

27. Base-line Measurement along the Ground.

Although base lines are sometimes measured along the ground, either after the surface has been carefully prepared by grading or levelling, or else after graded and levelled stakes have been put in at very close intervals—say 5 to 10 ft.—to support the band, the more general practice is to measure with the band supported in catenary. In the method of ground measurement, tension is applied at one end by means of a spring balance, and measurements are made to fine marks scratched on small flat sheets of brass or zinc fastened to the tops of pegs driven into the ground to the proper level at the end of every band length. This method is often used for the measurement of small local schemes of triangulation where geodetic accuracy is not needed and the full equipment for catenary measurements is not available.

28. Apparatus in Catenary Measurements.

In catenary measurements, the band is suspended in such a manner that the end marks lie close against marks carried on special measuring heads. Each of these measuring heads (fig. 2.15, Plate I) consists of a small metal spike or pillar, about 1 in. high and $\frac{1}{2}$ in. in diameter, mounted on a levelling head and centring arrangement by means of which the pillar can be made vertical and centred exactly over a ground mark, the whole being carried on a tripod similar to a theodolite tripod. One half of the circular top of the pillar is flat, and the index mark, to and from which end differences are measured, is cut as a very fine line across the centre of the flat surface. The other half of the top of the pillar is chamfered away slightly in opposite directions at right angles to the index mark. This chamfered surface is to allow for the slope of the band at its ends as it lies under tension with one edge against the line separating the flat surface carrying the index mark from the surface that has been chamfered away.

The small end differences between the end marks on the band and the marks on the measuring heads are generally read direct, either by means of fine graduations on the band itself at the zero and end marks or else by means of a suitable graduated metal measuring rule placed alongside the band. In fig. 2.15 a hinged magnifier, which enables graduations on the band and the index mark to be seen more clearly, is shown swung over to clear the pillar. Sometimes, however, special micrometers fitting on to the measuring head enable the differences to be read more exactly.

As corrections are required for the reduction of sloping lines to the horizontal, and for the reduction of the measured length of the base to its equivalent length at mean sea-level, the measuring heads are often provided with a special clinometer and with removable targets which fit on the pillar and can be used to observe the angle of slope between the end marks of each bay.

In practice, about six measuring heads are commonly used so that spares can be set up ahead of the measuring party and delays to the latter thus obviated. The method of measuring will be obvious. Commencing with the band suspended between measuring heads A and B, the measurements of the end differences give the uncorrected distance AB in terms of the nominal length of the band, and the measurement of the slope between the tops of the measuring heads gives data for the calculation of the slope and sea-level corrections. Temperatures, for temperature correction, are also read on two thermometers placed

PLATE I

Fig. 2.15.—MEASURING HEAD OF BASE LINE APPARATUS
(By courtesy of Messrs. Cooke, Troughton and Simms, Ltd.)

one near each end of the band. When these measurements are complete, the band is moved forward, suspended between measuring heads B and C and the previous measurements repeated, the measuring head A having meantime been removed and set up beyond any measuring heads already standing ahead of measuring head C.

29. Tensioning Apparatus.

In American practice, tension is often applied to one end of the band by means of a specially sensitive spring balance, the other end being fastened and adjusted at the proper height to a special straining pole or lever held by an assistant. The spring balance (fig. 2.16) is mounted on a ring which can be slid up and down a straining lever

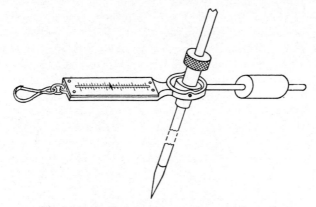

Fig. 2.16.—Spring balance mounted on straining pole

and clamped at any desired height, the balance being free to move about a horizontal axis on the ring and adjusted, by means of a counterweight, in such a way that it lies at an angle to the horizontal equal to the angle at which the end of the band will lie when the latter is under tension. The bottom end of the straining lever is held on the ground as fulcrum, and the top end is held by an assistant responsible for applying the correct tension.

In other parts of the world, including the British Colonies, tension is more usually applied at both ends of the band by weights supported by cords, which are fastened to the ends of the band and which pass over frictionless pulleys carried on special straining tripods (fig. 2.17, Plate II). These tripods are generally provided with two fine adjusting motions operated by two large wheels. One of these movements is at right angles to the plane of the pulley, and enables the latter and its mounting to be moved laterally so that the band can lie freely against the mark on the measuring head. The other movement enables the pulley to be moved up and down and adjusted to the correct height, the pulley itself being carried on a mounting which revolves freely about a vertical axis, thus enabling the pulley to adjust itself automatically to the vertical plane containing the band. For ease in handling, and to enable the tripod better to take up a forward thrust, the leg which comes below the line of the suspended band is made about twice the length of the other two.

30. Accuracy of Base Measurement.

Base lines are generally measured at least twice, once in the forward direction and once in the reverse direction, and in first-order geodetic work the difference between the two measures should not exceed one part in two millions (1/2M) of the length. This, however, should not be taken as a measure of the real error since a number of other, but unknown, errors, such as errors of standardization or temperature, exist but do not show up in the difference between the two measurements. It is very seldom, therefore, that a measured base will be accurate to more than about 1/500,000. In long chains of geodetic triangulation it is customary to measure check bases about 150 to 200 miles apart, and the difference between the length of a check base, as calculated through the triangulation from the original base, and the actual measured length seldom exceeds about 1/100,000.

INSTRUMENTS FOR THE ROUGH MEASUREMENT OF DISTANCE

31. Principal Methods.

The chain and steel band find their uses principally on surveys in connexion with the production of large-scale detailed plans, in measuring up work on the ground or in laying out work for constructional purposes. In all these cases a reasonable standard of accuracy is needed, and, as a result, the survey operations involved take some time to carry out. It sometimes happens, however, that accuracy, as the ordinary surveyor knows it, is not of great importance, and time for survey operations of a somewhat deliberate kind is not available. Cases like this arise in topographical mapping on a very small scale, or in exploratory or route surveys in which, compared with engineering or other surveys on a large scale, the work is very rough

PLATE II

Fig. 2.17.—STRAINING TRESTLE FOR BASE LINE MEASUREMENT

(By courtesy of Messrs. Cooke, Troughton and Simms, Ltd.)

and must be done very quickly. Hence, we now have to consider some instruments and equipment for the rough determination of distances by the following methods:

1. From observations of time intervals.
2. By pacing or pedometer.
3. By measurement with perambulator or cycle.
4. By measurements with ropes or wires.

32. Estimation of Distance from Observations of Intervals of Time.

This method consists in observing with a good watch the time taken to travel (by walking, horse riding, cycling, rowing, &c.) from one place to another. The rate of travel and the time taken being known, a very rough estimate of distance can be made.

The uncertain factor here is the rate of travel, which is not always constant and varies with the nature and slope of the ground. The ordinary rate can be found by taking the time interval required to travel over some measured distance such as that between two mile-posts.

33. Distance by Pacing or Pedometer.

Rough estimates of distance can be obtained by counting the number of paces required to cover the distance to be measured. Then, knowing the length of a pace, or the number of paces needed to cover a unit distance as obtained from pacing a known distance, the required distance can be estimated.

Instead of trying to keep count of paces, it is much more convenient to use a *pedometer, paceometer* or *passometer*. A *pedometer* (fig. 2.18) is a small instrument something like a watch in appearance. It should be carried attached to a point near the centre of the body, say from a waist-coat button. If it is attached to a leg, it will only register half the distance travelled or half the number of paces. The jolt at each pace causes a pointer to turn in the one direction. This pointer is worked

Fig. 2.18.—Pedometer
(By courtesy of Messrs. Hilger and Watts, Ltd.)

through a train of wheels operated by a pendulum which falls at each step, and is returned to its original position by a delicate spring. The movement of the pointer is read on the dial, which is usually graduated in miles and fractions of a mile. Pressure on a knob releases the pointer and lets it return to zero after a reading has been taken.

A pedometer should be tested over a known or measured distance. A special adjusting screw is generally provided by means of which the length of the stroke of the pendulum can be shortened or lengthened according as the recorded distance is too long or too short.

A *paceometer* or *passometer* is similar in appearance and operation to a pedometer but records paces instead of miles.

34. Distance by Measurement with Perambulator or Cyclometer.

A perambulator consists of an iron wheel of 6-ft. circumference attached to a frame carrying a handle, so that the wheel may be pushed along a road or path (fig. 2.19). As the wheel revolves, the number of revolutions are registered in terms of yards, furlongs and miles on an indicator contained in a brass dust-proof box mounted on the frame near the axle of the wheel.

Fig. 2.19.—Perambulator
(By courtesy of Messrs. W. F. Stanley & Co., Ltd.)

The perambulator works on exactly the same principle as an ordinary cyclometer or the speedometer on a motor car, either of which may be used for the rough measurement of distance along roads.

35. Distance Measurements by Rope.

The above methods of measuring distances more or less roughly necessitate a reasonably good surface over which to walk or travel, but distances have often to be measured over lanes or paths in forest, where the surface is anything but smooth and where there are many obstacles in the form of fallen trees, undergrowth, boulders, &c. For work of this kind, a rope is sometimes used. This rope is generally somewhat longer than the nominal length, the exact length chosen depending on the estimated average tortuosity of the path to be followed. Thus, a rope for a nominal length of 300 ft. may be 310 ft. long, the extra 10 ft. being an allowance for odd twists and turns in the line, a complete rope length being counted as 300 ft.

Ropes are inclined to vary a great deal in length, especially when new, or when there is a change in the dampness of the air or ground, the tendency being for them to shorten in length when they get wet. They should therefore be tested at intervals, preferably daily, against a chain or steel band, and, when they are new, it is well to season them before use by alternately wetting and drying them. They should always be pulled in one direction, so that the fibres are always being pressed lightly together.

CHAPTER III

INSTRUMENTS FOR MEASURING DIRECTIONS AND ANGLES—I

1. Introductory.

Although many small surveys in fairly open country can be completed entirely by means of a chain or steel band, using the ordinary methods of chain survey described in Chap. III of *Principles of Surveying*, the chain by itself is not sufficient when large areas are involved or when the survey is in heavily wooded or broken country. In such conditions, something more is needed, and it becomes necessary to use some sort of instrument which enables directions or angles either to be observed, or else to be plotted directly by an instrument such as the plane-table, or else determined indirectly from photographs taken in special cameras. In engineering practice, the direct observation of directions or angles by some form of compass or theodolite is far more commonly used, and in general is much more accurate than any system of replacing instrumental observation by graphical or photographic methods.

In addition to surveys of features already existing on the ground, much engineering work involves laying out definite directions or angles on the ground, and, as graphical methods are not suitable for such work, an instrument of the measuring kind is essential. The types of instrument to be considered in this chapter and the next may therefore be classified as follows:

(a) Instrument for the direct measurement of directions.
    1. Magnetic compass.
(b) Instruments for the measurement of angles.
    1. Sextant.
    2. Theodolite.
(c) Instrument for the measurement of either directions or angles.
    1. Miner's dial.

The *magnetic compass* has the advantage of giving readings directly in terms of directions or " bearings " referred to magnetic north, and is much simpler and quicker to use than the sextant or theodolite.

On the other hand, a compass is of limited accuracy as compared with either of the other two instruments, and magnetic bearings have the disadvantage that the direction of zero bearing—the magnetic meridian—not only varies from place to place as regards its orientation relative to the geographical north, or true meridian, but also at any one place it shows a slight variation during the day and in addition is subject to a slow, but appreciable, change from year to year. Another disadvantage is that the readings on a compass are affected by pieces of iron or other magnetic material in its immediate neighbourhood, and this may introduce serious error.

In spite of its comparatively low degree of accuracy, the magnetic compass has many uses in surveying. These naturally include cases where great accuracy is not needed, and, where this condition obtains, a complete survey may be made by chain and compass alone. In large-scale surveys, where the main angular work is measured by sextant or theodolite, a large compass mounted on a stand is often invaluable, by reasons of its portability, ease of handling, and the time that can be saved by its use, in the survey of unimportant or ill-defined detail, such as high- and low-water marks, the edges of lakes or streams, edges of forest land, paths and rides through woods, &c. There are also many cases where even small hand compasses can be used. These include all surveys of an exploratory or recon-naissance kind and different types of topographical surveys on small scales. In topographical work in forest country, in which an accurate and reasonably close network of control points is available, a stand compass is often used for the survey of main roads and for establishing a minor framework to which minor compass traverses, executed with hand compass and some comparatively rough method of chaining, can be tied and controlled. These minor compass traverses are used for such work as the survey of unimportant paths or streams, and for fixing the positions of spot heights on which to base the interpolation of contours.

The *sextant* measures angles, and not directions or bearings, and is a more accurate instrument than the compass, but not so accurate as the theodolite. It measures angles in any plane, whereas, when properly adjusted, the theodolite only measures angles in a horizontal or vertical plane. Hence, if an angle is observed with the sextant, and the true horizontal angle is required, the angles of slope to the observed points from the point of observation must be known or measured and the horizontal angle calculated from the observed angle and these angles of slope (p. 62).

The sextant, in general, is lighter, more portable, and quicker in use than the theodolite. One great advantage of it, however, is that it can be used from a moving platform, such as a ship or small boat, where it is impossible to use a theodolite: for this reason it forms a basic instrument in ordinary navigation, and it is employed extensively in hydrographic work to establish the positions of off-shore soundings taken from a small boat. In exploratory surveys on land it can be used for rough astronomical observations, and for measuring the angles of rough and rapidly executed triangulation schemes. In this last case, as there will usually be no very great differences in height between the observed stations, the observed angles will not differ greatly from the true horizontal angles, so that the error of treating them as horizontal angles will not be greatly out of proportion to the other errors inherent in rough triangulation schemes measured by sextant.

The *theodolite* is by far the most accurate instrument used by surveyors for the measurement of angles. Some theodolites are only provided with means of measuring horizontal angles, but more commonly they are also fitted with vertical arcs which enable vertical angles to be measured almost as easily as horizontal angles. A theodolite is always used mounted on a stand or tripod, or set on a concrete or brick observing pillar or other firm support. Different sizes are available, from the small theodolites, weighing only a few pounds, which are used for exploratory work, especially in mountainous country, to the very large instruments, weighing over a hundred pounds, which are used for geodetic work, and, of course, the larger the instrument the more accurate are the results to be obtained with it.* In most engineering work, a theodolite of intermediate weight, in which angles are read to the nearest minute or thirty seconds of arc, is employed both for actual survey observations and for setting out constructional works.

The theodolite is the king of survey instruments, as it can be used for more different and varied purposes than any other, although, possibly, the engineer's level is more extensively employed in ordinary engineering practice. But, in general land survey work, the theodolite is supreme.

In some theodolites, special *stadia hairs* are fitted which, used in conjunction with a graduated staff, enable distances to be deter-

---

* This is not universally true. Some of the modern theodolites with glass circles and optical reading devices are much more accurate than the older types of instrument of the same weight.

mined optically with an accuracy of about 1/100. In addition, a compass of some kind is often fitted to a theodolite, so that the instrument may be used to observe magnetic bearings as well as angles. The primary function of the theodolite, however, is to measure angles, and a compass is only an auxiliary fitting.

The *miner's dial* consists of a simple form of theodolite combined with a large surveyor's compass, and it can be used either as a theodolite or as a compass. When used as a compass to measure magnetic bearings it is little or no more accurate than a large surveyor's compass, but, by virtue of its horizontal graduated circle, it is capable of measuring differences of bearings, i.e. angles, more accurately than is possible with the compass and as accurately as is possible with a very small theodolite. As its name implies, it is chiefly employed in mining work.

A miner's dial is usually provided with a vertical circle, so enabling vertical as well as horizontal angles to be measured with it.

THE SURVEYOR'S COMPASS

2. General Description.

There are two main kinds of magnetic compass used in surveying —the surveyor's, or surveying, compass and the prismatic compass. The surveyor's compass is usually the larger and more accurate instrument, and is generally used on a stand or tripod. The prismatic compass is often a small instrument which is held in the hand for observing, and is therefore employed on the rougher classes of work.

The general principle of all magnetic compasses is very simple and depends on the well-known fact that, if a long, narrow strip of steel or iron is magnetized (i.e. made into a magnet), and is suitably suspended or pivoted about a point near its centre so that it can oscillate freely about a vertical axis, it will tend always to assume a certain constant direction, this direction being that of the *magnetic meridian* or *magnetic south-north* at the place where the observation is made. Hence, the surveyor's compass, shown diagrammatically in fig. 3.1, consists of a long, thin, pointed bar or needle of magnetized steel (1) with a small conical-shaped bearing (2) of agate or other very hard material at the centre. The end of this needle which points north, the north end, is differentiated from the other end, the south end, by a small metal pin (14) which passes horizontally through the needle near its north end. The agate bearing works on a pointed pivot (3) of hard steel carried at the centre of a low, cylindrical brass box (4). Attached to the opposite ends of a diameter of this box, or fastened to the base which supports it, are two sight vanes (5) and (6) which enable a definite line of sight to be defined or laid out. The brass box or case carries inside it, at the same height as the top of the magnetic needle, a fixed circular graduated arc (7) whose internal diameter is very slightly greater than the length of the needle. A disc of glass (12), fitting on top of the brass case, protects the needle and graduated circle.

In order to prevent unnecessary wear on the pivot and bearing, means are provided for lifting the needle off the pivot and holding

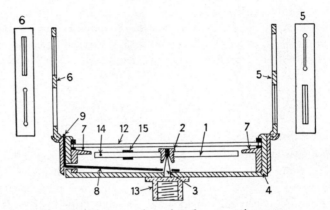

Fig. 3.1.—Diagrammatic section of a surveyor's compass

it clear while the instrument is not in use. Thus, in fig. 3.1, the left-hand vane when lowered presses down on a sliding piece (9), which, in turn, raises one end of the tongue (8), so lifting the needle from its pivot and holding it against the glass cover.

As the earth's magnetic field has a vertical as well as a horizontal component and this vertical component tends to pull down on one end of the needle as the instrument is moved from place to place, the strength of the pull varying with the place, most compasses are provided with a small metal rider (15) which can be slid along the needle so as to adjust it to lie in a horizontal plane.

In this instrument it will be noted that, when observations are being taken, the needle remains in a fixed position—the position of the magnetic south to north line—while the graduated circle, together with the line of sight, rotate about a vertical axis. Hence, the magnetic bearing observed is the angle between the direction which the line

of sight would have if it were made to coincide with the direction
of the needle and its direction when sighted on the object whose bearing
is required.   Note also that, if bearings are to be reckoned in one
direction from north, then the graduations of the circle must run in
the opposite direction.   This follows from the fact that, if the line of
sight is turned, say, to the right through an angle α from magnetic
north, then the reading on the circle will be the reading opposite the
needle, which in this case will lie at an angle α to the left of the zero
mark, which has moved with the line of sight.

While the above paragraphs describe the essential features of all
surveyor's compasses, individual instruments show considerable variations in design and in the additional fittings with which they are
provided.   In one form of the instrument, the circular box or case
containing the needle and the graduated arc is mounted at the middle
of a horizontal arm, about 12 in. long.   The sighting vanes are placed
at the two ends of the arm, not on the outside of the circular case,
and the arm also carries two spirit bubbles arranged at right angles
to one another.   This arm in turn is mounted on a levelling head
fitted with levelling screws, something similar to the levelling head
of a theodolite (p. 69), which enables the instrument to be set on a
tripod and levelled.   Sometimes, however, the ordinary levelling head
with screws is replaced by a ball-and-socket attachment which also
enables the compass to be levelled easily when it is set on a tripod.
In both these cases, the instrument can be rotated freely about the
vertical axis of the levelling head or ball-and-socket attachment, and
the sighting vanes on the arm are hinged so that they can be folded
down flat when the instrument is not in use.

In other cases, spirit bubbles are not provided, and levelling is
done simply by seeing that the needle swings in the same horizontal
plane as the plane of the graduated arc.   The long arm is omitted
and the sighting vanes are mounted, hinged to fold over the top of
the case, on the side of the latter as shown in fig. 3.1.   This type of
instrument is normally attached direct to some simple form of mounting
which enables it to be screwed on top of a tripod.

The size of a compass is defined in terms of the diameter of the
reading edge of the graduated arc.   The ordinary surveyor's compass
varies from about 2 in. to 8 in. diameter.

3. The Adjustments of the Surveyor's Compass.

The following are the adjustments usually necessary in the surveyor's compass:

(A) PERMANENT ADJUSTMENTS. (Adjustments which, once made, generally do not need repeating for a considerable time.)

1. Making the levels, when these are fitted, perpendicular to the vertical axis.
2. Making the sight vanes vertical when the instrument is levelled.
3. Adjusting the sensitivity of the compass.
4. Straightening the needle so that its ends lie in the same vertical plane with the pivot bearing.
5. Making the pivot coincide with the centre of the graduated arc.

(B) STATION OR TEMPORARY ADJUSTMENTS. (Adjustments which have to be made at every set-up of the instrument.)

1. Aligning and centring.
2. Levelling.
3. Focusing the prism (in the case of the prismatic compass only).

Permanent Adjustments

Making the Level Tubes Perpendicular to the Vertical Axis.—In instruments provided with level tubes, a line tangential to the bubble, when the latter is at the centre of its run, should be perpendicular to the direction of the vertical axis of rotation about which the instrument can be turned when the axis is vertical. For this purpose, the instrument is approximately levelled, and one bubble brought to the centre of its run when it is parallel to one pair of footscrews. The instrument is then rotated through 180° about the vertical axis. If the level tube is in adjustment, the bubble will remain in the centre of its run when the rotation is made. Otherwise, *one half* of the error must be corrected by means of the footscrews and the other half by the adjusting screws fitted to the level tube.

This adjustment will be more clearly understood from a study of the description of the adjustment of the level tubes of the theodolite given on pp. 99-101.

Making the Sighting Vanes Vertical when the Instrument is Levelled.—Set up a plumb bob and string a short distance away from the compass, or else take a well-defined vertical line on a nearby building, and, after the compass has been properly levelled, look at the string or line through the sight vanes. If the vertical hair in the one vane, or the edges of the slit in the other, are then seen not to be parallel to the string, remove the affected vane and file one side of the bed where it rests on the case or arm, or else insert a suitable packing, until a further test with the replaced vane shows that it is truly vertical when the compass is level.

Testing and Adjusting the Compass for Sensitivity.—After levelling the compass, lower the needle gently on its pivot and observe if it comes to rest quickly. If it does, and shows signs of sluggishness, the defect may be caused (1) by loss of magnetism, or (2) by wear of the pivot point.

A needle may be remagnetized by putting it in a solenoidal coil of insulated wire in which a strong direct electric current is flowing, but care has to be taken to see that it is put in the coil with its poles pointing in the same direction as the direction of the magnetic field formed by the electric current. This means that the north pole should be at the end of the coil which repels it when it is allowed to swing freely on the pivot. Otherwise, stroke the needle a number of times with a strong magnet, the strokes being made in such a direction that the north end of the magnet is drawn from the centre of the needle to the south end of the latter. Repeat the operation with the south pole of the magnet, working from the centre of the needle to its northern end.

The pivot can be sharpened by rubbing it on a fine oil-stone until it sticks when the point is touched by a finger-nail. During the grinding process the pivot should be kept rotating so that the grinding is systematically and symmetrically done about the pivot axis.

Straightening the Needle.—The needle may be bent in either a vertical or a horizontal direction. If bent in a vertical direction, any horizontal swinging of the needle when it is lowered on its pivot will be accompanied by a vertical seesaw motion of the ends. This motion ordinarily will not matter very much, but it may be overcome by taking the needle off the pivot and bending it, by means of a strong pair of pliers, in a vertical plane until the motion ceases.

To test for horizontal bending, revolve the instrument and note the readings at both ends of the needle for different positions of the zero of the graduated arc. If the difference of these readings is always some constant quantity other than 180°, the needle is bent but the pivot coincides with the centre of the graduations of the arc. If the difference is not constant, but varies with different positions of the zero mark, there may be an error of adjustment of both needle and pivot. In that case, the first thing to be done is to straighten the needle.

After the instrument has been properly levelled, read both ends of the needle in any position of the arc. Then revolve the whole compass until the mark originally against the north end of the needle comes against the south end. Now note the reading on the north end. If the instrument is in adjustment, this reading will be the same as the original reading at the south end, but if the instrument is not in adjustment, the difference in reading will represent approximately twice the error due to bending of the needle. If the difference exists, remove the needle from the pivot, and use a pair of strong pliers to bend the north end *half-way* towards the new position of the original reading at the south end. Replace the needle and repeat the test.

Adjustment of Pivot to Coincidence with the Centre of Graduations of the Arc.—Turn the instrument until the readings at both ends of the needle differ by exactly 180° and note this direction. Then turn the instrument through approximately 90° until the difference between 180° and the difference between the two end readings is a maximum. This difference represents the amount of the error caused by the pivot not being in coincidence with the centre of the graduations on the arc.

To adjust the error, take a pair of pliers and bend the pivot back in a direction perpendicular to the last direction of the needle and towards the segment of the arc which gave the larger angle when the instrument was turned to the second position.

This adjustment and the last should be repeated in their proper order, as they are both interdependent to some extent.

Temporary Adjustments

Aligning and Centring.—A compass, unlike a theodolite, does not need to be centred accurately over a station mark unless the bearings of more than one line, all radiating from the same point, are required. If the bearing of only one line is needed, the instrument can be set up at any point on the line, but if several bearings from the one point are involved, the compass must be set up as accurately as possible over that point.

The ordinary surveyor's compass has no fine centring device such as is almost invariably fitted to the ordinary engineering theodolite. Consequently, centring is done by manipulating the legs of the tripod until the instrument is seen to be centred over the ground mark or set on the proper alignment. A plumb bob can be used for centring over a mark, or, if a plumb bob is not available, a small stone or a coin can be dropped from the centre of the bottom of the tripod top so as to fall on the mark.

Levelling the Compass.—If the compass is not provided with level tubes, it must be levelled so that the needle rests in a horizontal position with both ends level with the graduated ring. If there are no levelling screws, or no ball-and-socket levelling attachment, this is done by suitable movement of the tripod legs during centring. If levelling screws and level tubes are fitted, the instrument is levelled by using the levelling screws in the manner described in connexion with levelling the theodolite (p. 109) to bring the bubbles in the tubes to the centres of their runs.

Focusing the Prism.—This adjustment is only necessary in the case of the prismatic compass (pp. 50–53). The prism is moved up and down in its slide at the side of the case until the graduations are seen to be sharp and in perfect focus.

4. Using the Surveyor's Compass.

The surveyor's compass is a very simple instrument to use. Having centred and levelled the instrument, release the needle to swing freely on its pivot and turn the graduated arc and the sighting vanes until the latter are in line with the object whose bearing is required. If the needle is very free and takes too long to settle down of its own accord, the lifting device which raises it off its pivot may be used as a brake to dampen down the larger oscillations. To do this lift it to touch the needle momentarily and very gently. When the needle has settled down, read the graduation opposite its north end, if necessary estimating fractions of a division. If the needle still persists in oscillating a little, observe the readings at the extreme limits, left and right, of a single swing and take the mean. This observation will give the magnetic bearing of the object. If accuracy is desired, similar readings should be taken at the south end of the needle, 180° being added to, or subtracted from, the result (when this is necessary) to give the equivalent bearing reckoned from north, the mean of the two observations being taken as the bearing of the line.

In all observations with a magnetic compass, it is important to see that there are no iron or steel objects near the instrument or on the person which would be likely to affect the needle. Steel rings in caps are sometimes a hidden source of disturbance.

The graduations on the surveyor's compass are usually designed either to give bearings reckoned from 0° to 360° measured clockwise from north, as in fig. 4.10a, or else they run both east and west from 0° at the north and south points to 90° at the east and west points, this system of reckoning being known as the *quadrantal* system because it runs in quadrants. The graduations are generally in terms of degrees, or of degrees and half-degrees, from which bearings may be interpolated by eye to thirds or sixths of a degree, i.e. to 20 or 10 minutes of arc.

THE PRISMATIC COMPASS

5. General Description.

Although the instrument just described is usually known as the surveyor's compass, the type known as the prismatic compass is also very generally used by surveyors. The so-called surveyor's compass is usually a fairly large instrument mounted on a tripod, but the prismatic compass may vary from a small hand instrument of about $1\frac{1}{2}$ in. diameter to a stand instrument of $4\frac{1}{2}$ in. diameter. A small pocket prismatic instrument of about $1\frac{1}{2}$ in. to 2 in. diameter is much used in the Army, and is often known as the " Service " compass.

The main features which differentiate a prismatic compass from an ordinary surveyor's compass are:

1. The graduations are either on a card or on a light aluminium annular ring fastened to the needle instead of being attached to the case carrying the sighting vanes, and the zero of the graduations coincides with the south point of the needle. The graduations therefore remain stationary with the needle, and the index turns with the sighting vanes. In the case of the surveyor's compass the graduations turn, and the index, which is the end of the needle, remains stationary.

2. Readings are taken through a magnifying prism attached to the outer case, and this prism reflects the magnified image of the graduations in a horizontal direction.

3. Readings are taken with reference to the lower part of a wire in the front sighting vane as index, instead of with reference to one end of the needle.

4. In order to give bearings reckoned clockwise from north, the graduations on the arc must run clockwise from the south end of the needle. In the surveyor's compass, it will be remembered, the graduations on the arc run anti-clockwise from north to give bearings reckoned clockwise from the same point.

PLATE III

Fig. 3.2.—LARGE PRISMATIC COMPASS

(By courtesy of Messrs. Hilger and Watts, Ltd.)

Fig. 3.3.—SERVICE PRISMATIC COMPASS
(By courtesy of Messrs. Hilger and Watts, Ltd.)

Fig. 3.4.—TROUGH COMPASS
(By courtesy of Messrs. Cooke, Troughton and Simms, Ltd.)

The prismatic attachment consists of a 45° reflecting prism with the eye and reading faces made slightly convex so as to magnify the image of the graduations. The prism is carried on a mounting which can be moved up and down between slides fixed on the outside of the case. The purpose of this up-and-down movement is to provide an adjustment for focusing. The image of the graduations is seen through a small circular aperture in the prism mounting, and immediately above this aperture is a narrow slit or a small V cut on top of the mounting, through or over which the vertical hair in the front vane may be viewed. The observation consists in noting on the graduated arc the reading which appears immediately underneath the vertical wire when the latter is directed on the station whose bearing is required. When the instrument is not in use, the prism and its mounting can be folded over about a hinge to lie flat against the side of the case of the instrument, and the front vane can also be folded about a hinge to lie flat over the lid of the case.

In the Service compass, the front vane is replaced by a folding lid containing a disc of clear glass, a vertical line etched on the glass replacing the vertical wire in the ordinary vane.

Fig. 3.2, Plate III, shows a prismatic compass for ordinary survey use, and fig. 3.3, Plate III, a small pocket Service prismatic compass. In fig. 3.2 the prism is in position ready for use, and in fig. 3.3 it is folded back in the position occupied when the instrument is not being used. The oblong mirror shown in fig. 3.2 in front of the forward vane slides up and down the vane, and is hinged to fold flat over it or to rest inclined at any angle with it. This mirror is used for solar observations, or for viewing any very high object, and is not a normal fitting to a compass. The two circular discs in front of the back vane are dark glasses which can be swung in front of the vane when solar observations are being taken. The small knob shown at the side near the prism in fig 3.3. is a *brake pin* for damping down the oscillations of the needle, and the other catch, the one on the right, is used for setting the instrument to read on a fixed bearing for night marching, or other similar operations, and is used in conjunction with the graduated rings on the outside of the case. Pressure on the brake pin causes a spring, in the form of a vertical strip of thin elastic metal, to press lightly against the edge of the graduated arc. The white arrow opposite the north point and the white line on the glass are luminous, and they, too, are used for night work.

In the form of prismatic compass known as the *liquid compass*, the needle and its accompanying graduated disc float in a special fluid contained in the case. The object of this arrangement is to reduce wear of the pivot and of the bearing of the needle. It also assists in damping down the needle's oscillations.

Most compasses, both surveyor's and prismatic, do not give absolutely true magnetic bearings, because almost every compass has its own individual " compass error ". The reason for this is that the magnetic and geometrical axes of the needle do not coincide exactly, and the result is that the end of the needle, or the zero of the graduated arc in the case of the prismatic compass, does not point exactly to the true magnetic north when the needle is swinging freely. The amount of the error is generally small, but may be appreciable. As this error is constant and of the same sign for all readings, and therefore affects all bearings equally, it ordinarily does not matter greatly. Its only effect is to cause a slight error in the orientation of the survey as a whole. The only means of determining its value is to observe the bearing of a line whose true magnetic bearing is already known.

6. Adjustments of the Prismatic Compass.

The adjustments of the prismatic compass are similar to those of the surveyor's compass already described, with the exception of the following:

1. There are no bubbles to adjust.
2. As a rule the sight vanes are not adjustable.
3. The needle cannot be straightened.
4. Levelling is done by holding the compass so that the graduated disc is swinging freely and appears to be level as judged by the top edge of the case.
5. There is the additional adjustment for focusing the prism which is described on p. 49.

7. Using the Prismatic Compass.

The prismatic compass is used in an almost exactly similar manner to the surveyor's compass, the only difference being that the readings are taken through the glass prism instead of direct on a graduated arc. If the instrument is a hand instrument, it must be held as steady as possible at eye level and pointed to the object whose bearing is required so that the object and the hair in the forward vane appear to coincide when viewed through the rear vane. The braking pin should be used if the needle does not settle down quickly enough, but care must be taken to ensure that the needle is properly free and the gradu-

ated disc reasonably level when the observation is being taken. When
the oscillations have become tolerably small, readings may be taken
in the prism at the limits of a single swing, and a mean value adopted.

TROUGH AND TUBULAR COMPASSES

8. Description and Use.

The trough and tubular compasses, unlike the surveyor's and
prismatic compasses, generally do not form a complete surveying instrument in themselves but are adjuncts of some other instrument,
usually a plane-table or theodolite in the case of the trough compass
or a theodolite in the case of the tubular compass.

The needle of the trough compass (fig. 3.4, Plate III) consists of a
long, narrow, magnetized bar of steel, pointed at both ends, with the
usual agate bearing at the centre. It is mounted in a narrow rectangular
box carrying a pivot at its centre, and this box is provided with a
tongue or lever, worked from a small stud or handle outside, which
enables the needle to be lifted off and kept clear of the pivot when the
instrument is not in use. At each end of the box, which is closed by a
glass cover, is a block of metal, the top of which is at the same level
as the tip of the needle, and on which is engraved a zero line and a
very short graduated arc extending about 5° on either side of the
central zero mark.

When the compass is used in conjunction with a plane-table, the
sides of the box are plane so that they can be used as a ruler. When
the compass is fitted on a theodolite, it is generally attached by screws
to the side of one standard, though sometimes it is fitted below the
lower plate.

The tubular compass consists of a magnetic needle mounted inside
a cylindrical metal tube. The north end of the needle carries a pointer,
and a circular glass disc, on which two fine, parallel, vertical lines are
etched, is fixed in the end of the tube facing the pointer. The setting
consists in lowering the needle on to its pivot and then rotating the
instrument carrying the compass until the pointer on the needle is
seen to be exactly midway between the vertical lines on the glass.
In some instruments, the glass is viewed with the naked eye, but in
others it is viewed through a magnifying eyepiece fitted at the end
of the tube.

The tubular compass, which is provided with the usual means of
raising the needle off its pivot, is attached by means of a special bracket

to the standards of the theodolite. It is the most accurate of the simpler forms of magnetic compass.

In a special form of tubular compass, manufactured by Messrs. Hilger and Watts, Ltd., and known as the " Conolly Standard Compass ", the tube, which is made of steel or other magnetic material and is itself magnetized, replaces the ordinary needle. This tube is carried in a stirrup suspended by a very light metal arm below an agate and pivot mounting. The tube is fitted with lenses so as to become a small telescope with a glass graticule, on which a fine scale is engraved, in front of the eyepiece. After the instrument has been levelled and the tube freed to oscillate, a ranging pole is ranged out in line with the central mark on the graticule scale. The tube is then taken out of its stirrup and replaced, after having been rotated through 180° about its longitudinal axis. If the magnetic and optical axes of the tube do not coincide, the ranging pole will no longer appear to be on the centre of the scale, the direction of the magnetic meridian then lying half-way between the zero mark and the reading of the apparent position of the pole on the scale. By this means, any error due to the non-coincidence of the magnetic and optical axes of the telescope can be eliminated. The instrument, however, as its name implies, is designed solely for standardizing other compasses, not for ordinary survey purposes.

It will be noted that the ordinary trough and tubular compasses only enable the direction of magnetic north to be determined: they do not in themselves enable magnetic bearings to be measured, this being done by the instrument to which they are auxiliary after the direction of the magnetic meridian has been determined.

CHAPTER IV

INSTRUMENTS FOR MEASURING DIRECTIONS AND ANGLES—II

THE SEXTANT

1. Principles of Operation and Description.

The operation of the sextant depends on bringing the image of one point, after suitable reflections in a couple of mirrors, into contact with the image of a second point which is viewed direct, the angle through which one mirror has to be turned to bring about this result being proportional to the angle between the two points.

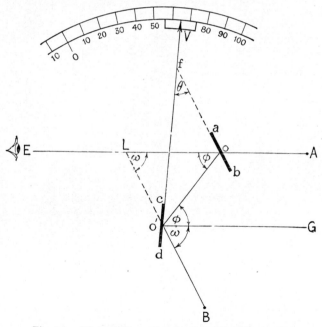

Fig. 4.1.—Diagram illustrating the principle of the sextant

In fig. 4.1 **aob** is a small fixed plate of glass, plain on the lower half and silvered on the upper, and **co′d** is another plate, silvered all

55

over on the surface facing **aob** and fixed on a pivot at **o'** so that it can
be rotated about this point as axis. The plate **aob** is called the *horizon
glass*, and the plate **co'd** the *index glass*. Let A be a point viewed directly
through the unsilvered part of **aob** by an eye placed at E. Let B be
a point whose image is reflected from the mirror **co'd** to **o** and thence
from the reflecting surface of the glass **aob** to E. Then, as the point
A is viewed directly along the ray E**o**, and B along a ray directly above
E**o** and parallel to it, the image of B will appear to the eye placed at
E to be superimposed above the image of A.

Produce **o'c** and **oa** to meet at **f** and the rays A**o** and B**o'** to meet
at L. Then the angle **o'**L**o** = ω is the angle between the points A and
B, and **o'fo** = θ is the angle between the mirrors. What we now require
to investigate is the relationship between ω and θ.

Through **o'** draw **o'**G parallel to **o**A and let angle E**oo'** = ϕ. Then
B**o'**G = **o'**LA = ω and G**o'o** = E**oo'** = ϕ. Also, by the laws of
reflection of light,

$$\mathbf{ao}E = \mathbf{o'ob} = 90° - \tfrac{1}{2}\phi,$$

and
$$\mathbf{co'o} = \mathbf{do'}B = 90° - \tfrac{1}{2}(\phi + \omega).$$

Hence, in triangle **o'of**,

$$\mathbf{o'of} = \phi + 90° - \tfrac{1}{2}\phi$$
$$= 90° + \tfrac{1}{2}\phi,$$

and
$$\theta = 180° - \mathbf{o'of} - \mathbf{co'o}$$
$$= 180° - 90° - \tfrac{1}{2}\phi - 90° + \tfrac{1}{2}(\phi + \omega).$$
$$\therefore \ \omega = 2\theta, \text{ or } \theta = \tfrac{1}{2}\omega.$$

Thus, for any one observation, the angle which the reflecting surfaces make with one another is half the angle between the objects
viewed.

In the actual instrument, the movable mirror **co'd** is mounted on
an arm, **o'**V, pivoted at **o'**, which carries an index mark and a vernier
at the end V. This arm, called the *index arm*, with its index mark
and vernier, moves along a fixed silver graduated arc centred at **o'**.
The graduations on the arc are usually numbered half their actual
angular values, so that the readings are directly in terms of ω, the
angle between the observed objects. The vernier (pp. 85–89) serves to
subdivide the readings on the scale so that observations can be taken
to 10″ or 20″ of arc. A telescope placed on the line of sight E**o** gives

PLATE IV

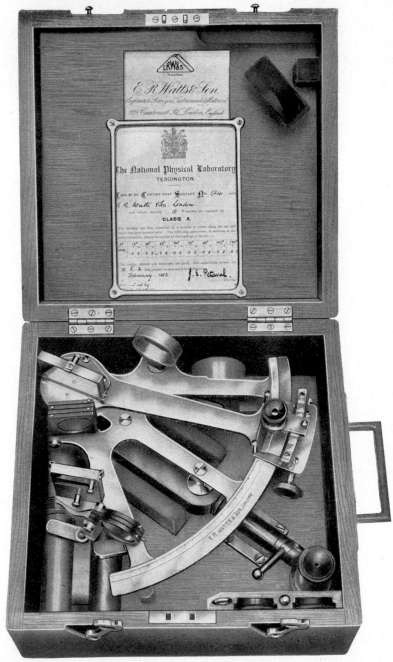

Fig. 4.2.—NAUTICAL SEXTANT
(By courtesy of Messrs. Hilger and Watts, Ltd.)

magnified images of the points A and B, and a small magnifying glass is fitted to enable the vernier to be read easily and clearly. Fine adjustments of the index arm and its attached mirror to obtain good contact of the images of the points observed can be made by means of special slow-motion screws, and other screws enable certain adjustments (described on pp. 58–59) of the mirrors to be affected.

There are three main types of sextant:

1. The nautical or astronomical sextant.
2. The sounding sextant.
3. The box sextant.

The *nautical sextant* (fig. 4.2, Plate IV) is specially designed for navigational and astronomical purposes, and is a fairly large instrument with a graduated silver arc of about 6 in. to 8 in. radius let into a gun-metal casting carrying the main parts. The arc is graduated to 20′ or 10′, but the vernier enables angles to be read to 20″ or 10″. The range of the scale is such as to enable angles up to about 150° to be measured. Two telescopes are usually provided, one for solar observations, and the other, of lower power and wider field, for terrestrial and stellar observations. Dark-coloured glasses, which can be swung into position when needed, are for use in solar observations. A clamp attached to the index arm enables the vernier to be clamped in any position to the graduated arc, and a tangential slow-motion screw provides the necessary fine-motion adjustment to bring the images into contact.

The *sounding sextant* is very similar in appearance and size to the nautical sextant but is not provided with dark glasses or shades. It is fitted with a specially large index glass to allow for the difficulty of sighting an object from a small rocking boat. The horizon glass has no transparent half, the direct sight being taken over the top of the glass. A telescope with a wide field is provided, but normally sights are taken through a peep-hole which can be fitted in the ring carrying the telescope.

The *box sextant* (fig 4.3, Plate V, p. 66) is much smaller and lighter than the nautical or sounding sextant, not much larger than a pocket instrument, and is most favoured by surveyors for ordinary land surveys. The graduated arc in this case has a radius of about 2 in. and is graduated to half-degrees from 0° to 120°, the vernier enabling angles to be read to single minutes. The working parts of the instrument are contained in a circular brass case of about 3 in. to 4 in. diameter, the index arm and mirrors being protected when the instrument is not in use by a lid which can be removed and screwed on the bottom of the case to form a handle when observations are being taken. The larger of the two milled heads shown in the illustration operates a pinion, which engages with a toothed circular segment carrying the index glass and index arm. By this means a controlled fine motion can be given to both index glass and index arm. The smaller milled head is a key which can be used to operate adjusting screws for eliminating index error and for adjusting the horizon glass. A small removable telescope may be used for long sights, but ordinary observations are generally taken through a peephole in the case. Dark glasses are also fitted, but the object of these is to enable the sun to be used as the very distant object when adjusting the instrument for index error rather than to enable the sextant to be employed for ordinary solar observations.

2. Adjustment of the Nautical and Sounding Sextants.

There are four adjustments possible and necessary in the ordinary nautical and sounding sextants. These are:

1. The index glass must be made perpendicular to the plane containing the graduated arc.
2. The horizon glass must also be made perpendicular to the plane containing the graduated arc.
3. The line of sight must be made parallel to the plane of the graduated arc.
4. The mirrors should be made parallel to each other when the index on the vernier is at zero. This means that the direct and reflected images of a single distant object should coincide when the reading on the arc is zero.

All of the above adjustments are under the control of, and must be made by, the surveyor himself. In addition to them, however, there are three special conditions that depend on adjustments made or work done by the maker during the process of manufacture and which must be fulfilled. They are:

1. The axis of rotation of the index arm should coincide with the centre of graduations of the arc.
2. The pieces of glass forming mirrors and shades should have their opposite faces truly plane and parallel.
3. The graduations of the arc should be uniformly and accurately spaced.

The effects of the non-fulfilment of any one of these conditions can be allowed for by applying corrections to all observed readings, and the three corrections can also be combined to form a single correction which will vary for different positions of the index arm. Accordingly, when accuracy is desired, the sextant should be tested at the National Physical Laboratory, or similar institution, and a table prepared giving the magnitude and sign of the combined correction for different settings on the graduated arc. Very often such a table is provided by the makers with the better classes of instrument.

The maker is also generally responsible for the adjustment of the index glass and telescope of the box sextant, and in this instrument no means for enabling the surveyor to make these adjustments himself are provided. Accordingly, the only adjustments possible in this case are those which involve the horizon glass.

Adjustment of Index Glass.—To make the index glass perpendicular to the plane of the arc, clamp the index arm firmly at the middle reading of the scale, and, looking into the glass where it is level with the eye, see if the reflected image of the scale is a continuation of a part of the arc itself as seen directly. If not, adjust by means of the screw at the back of the mirror so as to slant the glass forwards or backwards according as the reflected image appears to fall or rise from the part of the arc that is viewed directly.

Adjustment of Horizon Glass.—This adjustment means making the plane of the horizon glass perpendicular to the plane containing the graduated arc. For this purpose, sight a star and move the index arm slowly backwards and forwards on either side of the zero mark. If the reflected image of the star appears to pass directly over the direct image, the horizon glass is in adjustment. If not, turn the small screw at the top of the horizon glass until the test is satisfied.

In the case of the sounding sextant, the test is made on a very distant straight line, the direct and reflected images being made to appear to form a single straight line as the sextant is slowly tilted.

Adjustment of Line of Sight of Telescope.—The object of this adjustment is to make the line of sight of the telescope parallel to the plane of the graduated arc. The long, or inverting, telescope supplied with the nautical sextant is generally provided with a graticule consisting of four lines forming a square. Set the telescope so that two wires are approximately parallel to the graduated arc. Sight a star directly and bring the reflected image of a second star, at least 90° distant from the other, to coincide with the first image on one wire. Move the instrument slightly until the images appear on the second wire. If these images still appear to be in contact, the adjustment is satisfactory; if the test is not satisfied, adjust the inclination of the telescope by means of the adjusting screws on the collar into which it is screwed.

This adjustment is not necessary in the case of the sounding sextant as the telescope on it is normally not adjustable.

Adjustment for Index Error.—Index error represents an error in reading the angle due to the fact that the index on the vernier is not at zero when the observed angle is zero. Since this error is inclined to vary under different conditions of temperature, it is usual to determine its magnitude and apply this as a constant correction to the observed angles rather than to attempt any adjustment.

From fig. 4.1 it will be obvious that, if an object is very distant, Ao and Go′ may be considered to be rays from it which, because of the great distance involved, are parallel rays. In that case, the angle ω is zero, the mirrors **aob** and **co′d** are parallel to one another, and the vernier should be at zero on the graduated arc. The test therefore consists in pointing the telescope at a star or very distant object and bringing the reflected image of the same star or object into contact with the direct image. The vernier should then read zero, but if it does not, the amount by which the index on it lies to one side of the zero on the scale is the index error. If the vernier index lies off the main scale, the error is positive and the amount of it must be added to the observed angles; if it lies on the main scale, the error is negative and its amount must be subtracted from the scale readings.

For purposes of determining this error, the graduations on the arc are generally extended a short way to the negative side of the zero mark.

Since index error represents an amount which has to be applied to all readings, it should be determined very carefully, several determinations being made at one time and a mean result adopted. In addition, it is well to determine it at frequent intervals, especially when there is a considerable variation in temperature since the last determination.

If it is desired to adjust for this error, clamp the vernier at zero, and adjust the horizon glass by means of the adjusting screw at the bottom of the frame containing the glass until the direct and reflected images of a star or very distant body are in coincidence. This adjustment may upset the adjustment for perpendicularity of the horizon glass, so that this last adjustment should be re-tested and, if necessary, re-made. If it has to be re-made, a further adjustment for index error may be needed.

3. Adjustment of the Box Sextant.

In the box sextant no provision is made for the adjustment of the index glass and telescope, but the horizon glass is fitted with the necessary screws to permit of the glass being adjusted for perpendicularity and for index error. The tests can be made as described above. The objects used in both tests should be very distant, or else the sun may be used, dark shades being provided for this purpose. When the sun is used for testing index error, the direct and reflected images of both limbs should be brought into coincidence when the vernier reads zero, the combined images then forming a circular disc with no breaks at the limbs.

4. Using the Sextant.

As the sextant normally is used in the hand, it must be held as steady as possible and pointed at one station; holding the instrument against an upright ranging rod will help to steady it. The index arm is then moved until the reflected image of the second point is brought into the field of view close to the direct image of the first point. The index arm is now clamped, and the two images brought into exact contact by means of the tangent fine-motion screw. The reading on the arc is then taken by means of the vernier and the magnifier provided with the instrument.

If the box sextant is used, the index arm is moved, and the images brought into contact by means of the milled head which actuates index glass and arm, after which the vernier is read as before.

If an angle greater than the range of the instrument has to be measured, or if the reflected image is not sufficiently bright because of the large angle being observed, the measurement can be made in two parts; first to some suitable intermediate point lying or set out between the two points to be observed, and then from this intermediate point to the second main station. This intermediate point should lie very approximately in the plane containing the two observed stations and the station of observation.

5. Observing Vertical Angles with the Sextant: The Artificial Horizon.

If vertical angles of elevation have to be measured with the sextant, it is necessary to employ an artificial horizon. This is a shallow tray containing mercury. The angle between a distant point as viewed directly and its reflected image seen in a pool of mercury is equal to twice the angle of elevation of the point. Fig. 4.4 will make this clear. Since the point is a distant one, the rays AC and BD from it are parallel. By the laws of reflection of light, angle BDE = CDF, and this is equal to DCL, where CL is a horizontal line through C, C being the point where the sextant is held. But ACL and BDE are equal as AC and BD are parallel. Hence, angle ACD, which is the angle measured with the sextant, is twice the angle ACL, the angle of elevation of the point.

The artificial horizon is used in astronomical observations with the sextant on land, but it cannot normally be used at sea because of the motion of the mercury surface. In navigation, therefore, angles of elevation of the sun are obtained by observing the vertical angle between the horizon of the sea and the sun, a correction, depending on the height of the sextant above sea level, being applied to correct the observed angle for the angle of *dip* between the observer and the visible horizon.

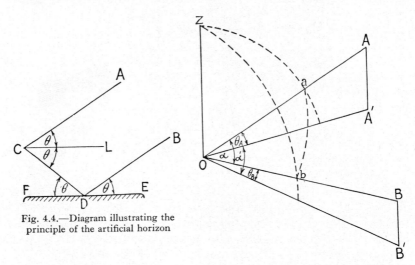

Fig. 4.4.—Diagram illustrating the principle of the artificial horizon

Fig. 4.5.—Reduction to the horizontal of the angle measured with the sextant

6. Reduction of Angles observed with the Sextant to the Horizontal Plane.

In fig. 4.5, let O be the point of observation and A and B the stations sighted. Let A′ and B′ be the projections of A and B on the horizontal plane through O. Then the angle observed with the sextant is the angle AOB = α, and the horizontal angle, which is now the angle required, is the angle A'OB' = α'.

Let AOA' = θ_A be the slope or inclination to the horizontal of the line OA and BOB' = θ_B be the slope or inclination to the horizontal of the line OB.

Put
$$z_A = 90° - \theta_A,$$
$$z_B = 90° - \theta_B,$$
$$s = \tfrac{1}{2}(\alpha + z_A + z_B).$$

Then
$$\tan \tfrac{1}{2}\alpha' = \left[\frac{\sin(s - z_A)\sin(s - z_B)}{\sin s \sin(s - \alpha)} \right]^{\frac{1}{2}}.$$

This is the relation required. The reader who is familiar with spherical trigonometry will recognize that this formula can easily be derived by imagining a sphere drawn round O, as shown dotted in fig. 4.5, OZ being the vertical through O and **a** and **b** the points where the lines OA and OB are cut by this sphere. Then, in the spherical triangle **Zab**, the sides **Za**, **Zb** and **ba** are z_A, z_B and α respectively and the angle **aZb** is the angle α', which is required. Hence, the three sides of the triangle being known, the angle α' can be calculated.

THE THEODOLITE

7. Types of Theodolite.

Up to some years ago, there were two main types of theodolites in common use, one known as the *wye* theodolite and the other the *transit* theodolite. The wye theodolite has now become practically obsolete, and specimens are only to be found in old-fashioned offices or in shops selling second-hand instruments. Consequently, in what follows we shall confine ourselves mainly to a consideration of the transit theodolite, contenting ourselves merely with a very brief description of the points in which the wye theodolite differs from the transit instrument. If further information about the wye theodolite, its use and adjustments, is desired, reference should be made to one of the larger textbooks.

In England the transit theodolite is simply called a *theodolite*, it being understood that a theodolite of the transit type is meant, but in Canada and the United States a theodolite is almost invariably called a *transit*, it again being understood that a transit theodolite is meant. The term *transit* is open to the objection that another, older, instrument called a transit is commonly used in astronomical observatories for observing transits of stars over the cross hairs in the telescope.

8. Description of Transit Theodolite.

A theodolite consists of the following essential parts (see fig. 4.6) in addition to the tripod:

1. A *levelling head* which supports the main working parts of the instrument and which screws on to a tripod. The head comprises two parts: (*a*) a *levelling base* or *tribrach* and *trivet plate* (1) fitted with levelling footscrews (2) for levelling the instrument, i.e. for making the vertical axis vertical, and (*b*) a movable head or centring arrangement (3) for centring the vertical axis accurately over a ground mark or under a mark overhead, as on the roof of a tunnel.

2. A lower circular horizontal metal plate (4) carrying a circular graduated arc (5). This *lower plate* is attached to a vertical metal spindle which works in a vertical bearing forming part of the levelling head.

3. The centre of the vertical spindle of the lower plate is bored out to form a bearing for another vertical spindle which carries an upper circular horizontal plate. This *upper plate* (6) carries: (*a*) an index and vernier (10) or micrometer enabling fine readings to be taken on the graduated horizontal arc, (*b*) standards (7) for supporting a telescope, and a spirit level (15) (in some cases two) for indicating when the instrument is level.

4. A telescope (8) fitted to a *horizontal, trunnion,* or *transit axis* (9) supported in bearings carried on the standards of the upper plate.

These are the main essential parts for the measurement of horizontal angles, but most theodolites are designed to enable vertical angles also to be measured. For this purpose they are fitted with

5. A circular graduated arc carried on a *vertical circle* (11) attached to the horizontal axis of the telescope.

6. A *vernier frame* or arm (13) carrying an index and verniers or micrometers to enable vertical angles to be measured.

Fig. 4.6 shows the essential main parts diagrammatically but, in addition to the main parts, the following are also essential features of a theodolite:

1. A *plumb bob* which can be attached to the bottom of the vertical axis to indicate when the instrument is central over a ground mark.

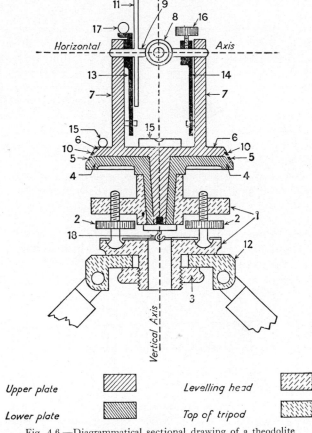

Fig. 4.6.—Diagrammatical sectional drawing of a theodolite

1. Tribrach and Trivet.
2. Footscrews.
3. Clamping Screw for centring.
4. Lower plate.
5. Graduated arc.
6. Upper plate
7. Standards.
8. Telescope.
9. Horizontal axis.
10. Verniers.
11. Vertical circle.
12. Tripod top.
13. Vernier frame.
14. Arm of vertical circle clamp.
15. Plate levels.
16. Vertical circle. Clamping screw.
17. Level on vernier arm.
18. Hook for plumb bob.

2. A *lower clamp* and *lower tangent screw* (see fig. 4.7, p. 66), the one for clamping the lower plate and its hollow vertical axis to the levelling base, and the other for enabling a finely controlled circular motion with reference to the base to be given to the lower plate about the vertical axis.

3. An *upper clamp* and *upper tangent screw* for clamping the upper plate to the lower one, and enabling a finely controlled circular motion about a vertical axis to be given to the upper plate relative to the lower plate.

4. A *graticule*, with *cross hairs*, in the telescope to give a definite line of sight.

5. A clamp and fine-motion screw for clamping the vertical circle and enabling a finely controlled circular motion to be given to the combined telescope and vertical circle about the horizontal axis.

6. A reading device, either vernier or micrometer, fitted to the vernier frame to enable fine readings on the vertical circle to be taken. Usually two verniers or micrometers, at opposite ends of a diameter, are fitted for this purpose. The vernier frame usually has a clip and fine-motion screw, working between a lug on the standards, to enable the position of the frame to be adjusted slightly relative to the standards.

7. In certain cases a level tube is fitted either on top of the vernier frame or on top of the telescope. Occasionally, a level tube is fitted in both places.

The last two parts mentioned [i.e. (6) and (7) above] are only fitted on theodolites which are designed to measure vertical as well as horizontal angles.

There are three imaginary geometrical lines, two of which are axes of rotation and one an optical axis, which are of importance. These are:

1. The *vertical axis*. This is the vertical axis about which the lower and upper plates rotate. Each plate has its own axis but the two should coincide.

2. The *horizontal axis*. This is the axis about which the telescope and vertical circle rotate.

3. The *line of sight* or *line of collimation*. This is a definite line of sight passing through the intersection of the horizontal and vertical cross hairs in the telescope and the optical centre of the object glass of the telescope. This line should be perpendicular to the horizontal axis, and it should also be truly horizontal when the reading on the vertical circle is zero and the bubble on the telescope or on the vernier frame is at the centre of its run.

In a perfectly adjusted theodolite these three lines would meet in a point.

As a preliminary understanding to the working of the theodolite, it is very important to form a clear idea of the action of the clamps and tangential screws of the lower and upper plates, which we may tabulate as follows:

1. Both clamps tightened. Both lower and upper plates, complete with telescope and vertical circle, are fixed with reference to the

PLATE V

Fig. 4.3.—BOX SEXTANT
(By courtesy of Messrs. Hilger and Watts, Ltd.)

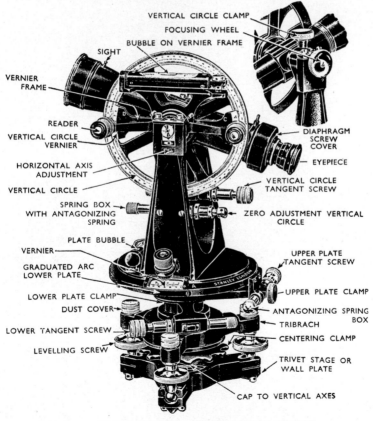

VERTICAL CIRCLE CLAMP
FOCUSING WHEEL
BUBBLE ON VERNIER FRAME
SIGHT
VERNIER FRAME
READER
VERTICAL CIRCLE VERNIER
HORIZONTAL AXIS ADJUSTMENT
VERTICAL CIRCLE
SPRING BOX WITH ANTAGONIZING SPRING
PLATE BUBBLE
VERNIER
GRADUATED ARC LOWER PLATE
LOWER PLATE CLAMP
DUST COVER
LOWER TANGENT SCREW
LEVELLING SCREW

DIAPHRAGM SCREW COVER
EYEPIECE
VERTICAL CIRCLE TANGENT SCREW
ZERO ADJUSTMENT VERTICAL CIRCLE
UPPER PLATE TANGENT SCREW
UPPER PLATE CLAMP
ANTAGONIZING SPRING BOX
TRIBRACH
CENTERING CLAMP
TRIVET STAGE OR WALL PLATE
CAP TO VERTICAL AXES

Fig. 4.7.—SIMPLE VERNIER THEODOLITE BY MESSRS.
W. F. STANLEY & CO., LTD.
(By courtesy of the makers)

vertical axis and levelling base. Fine motions are possible. If the lower tangent screw is turned, both lower and upper plates rotate together as a single unit about the vertical axis with reference to the levelling head. If the upper tangent screw is turned, the upper plate rotates about the vertical axis relatively to the lower plate.

2. Lower clamp loosened, upper plate clamped. Both lower and upper plates may rotate as a single unit about the vertical axis with reference to the levelling head. There is no point in operating either tangent screw.

3. Lower plate clamped, upper plate loosened. Upper plate, complete with telescope, etc., can be rotated about the vertical axis relatively to the lower plate. Operating the lower tangent screw will turn the lower plate relative to the levelling head, leaving upper plate loose and free to turn.

4. Both clamps loosened. Both plates are free and can be rotated independently about the vertical axis. Tangent screws not in use.

9. Setting and Reading Procedure for Horizontal and Vertical Angles.

Let us now consider the procedure involved in setting and reading the theodolite so as to give a direct observation of a horizontal angle between two stations, it being assumed that the instrument has been properly centred, levelled and focused, and that it is in perfect adjustment.

1. Clamp lower plate in any position by means of the lower clamp and, with the upper clamp loose, rotate the upper plate until the index on it is very approximately at the zero point on the graduated arc. Tighten the upper clamp and use the upper fine-motion tangent screw to bring the index into exact coincidence with the zero mark.

2. Loosen the lower clamp and turn the instrument until the line of sight is directed approximately at the first station. Tighten the lower clamp and bring the image of the first station into coincidence with the cross hairs of the telescope by means of the lower tangent screw. The lower plate is now fixed in position so that the reading on the arc is zero when the line of sight is directed to the first station.

3. Keeping the lower clamp fixed, loosen the upper clamp and turn the instrument so that the line of sight appears almost to intersect the second station. Tighten the upper clamp and bring the line of sight into exact coincidence with the image of the second station by means of the upper tangent screw. During this operation the lower plate with its graduated arc has remained fixed throughout, and the reading on it was zero when the line of sight intersected the first station. The upper plate has meantime moved through the angle necessary to bring the line of sight from the first station to

the second station, and the index has moved with it. Consequently, if the vernier is now read, the angle measured on the lower plate is the horizontal angle required.

If the vertical circle on the instrument was on the observer's right during this measurement, the position of the instrument is known as *face right* or *circle right*. If the telescope is now turned end for end about the horizontal axis, the upper clamp released and the upper part of the instrument turned through 180°, the telescope will point towards the last station, but the vertical circle will now be on the observer's left. This position of the instrument, with the vertical circle on the observer's left, is known as *face left* or *circle left*. If desired, the angle can be measured again with the circle in this new position, but, if there are two verniers, the one which previously read 0° when the telescope was directed to the first station will now read 180°, and the one which previously read 180° will now read 0°. If an angle is measured twice, once face right and once face left, and the mean of the two results taken, the result will give a measure of the angle from which the effects of an instrumental error, known as *collimation error*, have been eliminated.

Again, if the instrument was turned clockwise from the first station to the second when the original measurement was being taken, the movement is known as *swing right*, but if it had been turned in the opposite direction, the movement would be known as *swing left*. Usually a swing in one direction should be balanced by a swing in the opposite direction when a second observation is being taken on a different face. Taking the mean of two observations when one is observed swing right and the other swing left, has the tendency to eliminate the effects of errors due to friction or backlash in the moving parts.

The measurement of vertical angles is simpler in principle. If the instrument is again assumed to be in perfect adjustment, so that the bubble on the vernier frame or on the telescope (if one is fitted) is perpendicular to the vertical axis when the bubble is at the centre of its run, and the reading on the vertical circle is zero when the line of collimation is truly horizontal, the procedure is as follows:

1. Level the instrument carefully, using the level on the vernier frame or on the telescope if either is fitted, and then loosen all clamps including the clamp of the vertical circle.
2. Point the telescope to the station to which the vertical angle is to be measured, and, by means of the clamp of the vertical circle, clamp when the line of sight intersects the station roughly. Then

use the fine-motion tangent screw of the vertical circle to bring the image of the station exactly on to the middle of the horizontal hair in the graticule. The reading of the vernier or micrometer is now the angle required or its complement.

If an accurate value is required, the observation should be repeated on the other face to eliminate collimation error.

It should be noted that, in measuring a horizontal angle, the graduated arc remains fixed while the index and verniers rotate. In measuring a vertical angle, the index and verniers remain fixed while the graduated circle rotates.

The above is a general description of the basic principles involved in the construction and use of all theodolites, but considerable variations exist in the general design and features of instruments made by different makers, or even of different models produced by the same maker. It is impossible to describe here all the variations that exist, but, in the meantime, we proceed to a detailed study of the different parts and fittings that go to make up a typical theodolite, most of which are mentioned briefly above. An illustration of an ordinary simple theodolite is given in fig. 4.7, Plate V, p. 66, and the student will do well to study this illustration in conjunction with the sectional diagram in fig. 4.6, p. 65, when reading the following pages.

10. Levelling Head.

The function of the levelling head is to support the main part of the instrument, to attach it to the tripod, and to provide a means for levelling it. As a matter of convenience, some sort of centring device, by which the instrument can be accurately and easily set over a mark on the ground or under an overhead mark, is usually added.

In some instruments, particularly in some of the older models, there are four levelling screws at 90° intervals, but in recent designs the general practice is to provide three only, 120° apart. In theory, the three-screw arrangement is the better one, as three points of support are sufficient for stability and the introduction of an extra point of support leads to uneven wear on the screws and may cause strain in the latter and in the two main axes. On the other hand, the four-screw arrangement makes it possible to use a somewhat simpler and lighter levelling head and is sometimes favoured on that account.

In a three-screw arrangement, the levelling screws, which are of the form shown in fig. 4.7, are carried in a special casting called a *tribrach,* which is provided with three arms, 120° apart, and with a central cylinder hollowed out to form a long cylindrical or slightly conical bearing for the spindle of the lower circle. The screws work in threaded holes near the ends of the three arms, and their bottom ends, which are generally slightly enlarged to form hemispherically shaped feet, rest in recesses in a plate called a *trivet stage, wall plate,* or *tribrach foot plate.* This trivet stage is a casting having a large-diameter hole in the centre, threaded at the bottom, which screws on to a metal casting fastened to the top of the tripod and forming part of the tripod head, or on to a plate forming part of a centring device.

In some theodolites the trivet stage is separate and detachable from the main tribrach, and is provided with a top plate in which

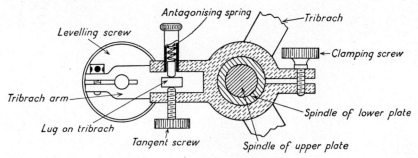

Fig. 4.8.—Clamp and tangent screw of the lower plate of a theodolite

there are three pear-shaped holes. The plate, which is rotatable about its centre, is first placed with the largest parts of the holes over the recesses in the bottom plate. The feet of the levelling screws are then passed through these holes so as to rest in the recesses. The top plate can now be rotated to bring the smaller parts of the holes over the tops of the feet of the levelling screws and can then be clamped by a screw provided for the purpose. In this way, the tribrach and the upper part of the theodolite are firmly fixed to the trivet stage.

In most theodolites the trivet stage is not detachable from the tribrach and the two form a single unit, the feet of the levelling screws being held in position in the recesses shaped to take them.

It is very important that there should be no play between the levelling screws and the tribrach arms which hold them. Accordingly, provision is almost invariably made to enable any wear in the screws to be taken up. There are many devices for doing this. In the simplest form, the ends of the arms of the tribrach have a fairly wide vertical slot between the hole for the screw and the outside of the arm, and small horizontal adjusting screws, working through one face of the slot and passing through the other, enable the slots to be slightly closed or opened, so making it possible for the pressure on the sides of the screws to be adjusted at will (fig. 4.8). In many models the tops of the levelling screws, where they project above the tribrach arms, are provided with dust caps to protect the screw and keep it free from dust.

When a centring device or movable head is fitted to a theodolite it is sometimes placed immediately below the trivet stage, but sometimes it is above the tribrach, between it and the lower plate. The latter arrangement has the advantage that centring may be done after the instrument has been levelled, and is therefore not likely to be disturbed by any subsequent levelling. One form of movable head consists of two plates with large circular apertures in their centres, the lower of which is supported on the flat surface of the casting forming the tripod head. This lower plate is pivoted near one edge of the casting and carries a threaded pin, which works in a curved slot near the opposite edge of the tripod casting. A butterfly nut on the pin serves to clamp the plate to the casting when required. On top of this lower plate is another plate pivoted to it 90° from the first pivot and fitted with another threaded pin, complete with butterfly nut, which works along a curved slot in the lower plate. This upper plate carries at its centre a threaded collar. The trivet stage on which the theodolite is supported can be screwed on to this collar. Unclamping the lower plate allows this plate to be rotated about its pivot through a fairly large angle, and unclamping the upper plate allows the latter to be rotated in a direction at right angles to the first. In this way, the threaded collar supporting the trivet stage and theodolite can be given a considerable lateral motion in two directions at right angles to one another, and so a plumb bob attached to the vertical axis of the theodolite can be brought easily and quickly over a station mark. The great advantage of this type of movable head is that it permits a considerable range of movement.

Another form of movable head, which is simpler than the one just described but does not allow such a large range of movement, consists of two plates with flat surfaces permitting the one to slide easily and smoothly over the other. The lower plate is part of the casting to which the tripod legs are attached and has a large circular aperture in the centre. The upper plate is the underneath part of the trivet stage carrying the theodolite and can be clamped as desired to the tripod head casting by a large, threaded, hand clamping ring underneath the casting. This ring works on a long, hollow, threaded central cylinder that forms part of the trivet stage, and projects below the bottom of the tripod head casting. The centring device depicted diagrammatically in fig. 4.6 (p. 65) is of this kind.

Much the same principle is used when the centring device is above the tribrach. In this case, the top of the tribrach is ground to a flat surface, and a flat circular plate carrying the bearing for the spindle of the lower graduated plate slides over the top of the tribrach and can be clamped when the plumb bob is over the station mark.

11. Lower Plate and Graduated Arc.

The lower plate carrying the graduated arc consists of a casting in the form of a circular plate with a central spindle at right angles to the plate. The outer surface of the lower part of the spindle may be cylindrical, but often it is in the form of the frustum of an elongated cone with the larger end immediately below the plate. This outer surface is ground to a smooth polished surface to fit snugly into the vertical bearing in the tribrach, this bearing also being ground to a smooth cylindrical or conical surface to receive the spindle. A short length of the top of the spindle, near its junction with the plate, is cylindrical in shape to take a split collar. At one side this collar carries a clamping screw by means of which it can be slightly loosened or tightened, and at the other side a tangent fine-motion screw and a spring buffer, or *antagonising spring*, working on either side of a lug on top of the tribrach. The object of the clamp and tangent screw, which are shown in fig. 4.8 (p. 70), is, of course, to enable the vertical spindle to be clamped to the tribrach, and a fine motion relative to the latter to be given to it by means of the tangent screw. The centre of the spindle itself is hollowed out to form a bearing for the spindle of the upper plate.

The above is a description of about the simplest form of clamp and fine-motion fitting which it is possible to have, but many other designs, each with its own special advantages, are possible. Fig. 4.9 for instance, shows the clamping and fine-motion arrangements for both lower and upper circles which are fitted by Messrs. Cooke, Troughton & Simms to some of their theodolites.

In the ordinary theodolite the *graduated arc* consists of a silver ring on which the graduations are engraved. This ring is inset into the lower horizontal plate or is permanently fixed to it near or at the rim. On most theodolites, the outer rim of the lower circle is not flat, but is bevelled off from the horizontal and the silver graduated ring lies on the bevelled surface, the object being to provide a con-

veniently placed surface for observing. In other instruments the ring
may be flat on a flat lower plate, or it may be vertically placed on the
outer vertical rim of the plate.

During the last two decades there has been a tendency in the
better classes of theodolites to replace graduations on silver by graduations on glass, the upper part of the lower plate then consisting of a
glass disc on which the graduations are etched. This arrangement
is almost always, if not always, combined with a system of prisms
and lenses by means of which an image of the graduations, near the

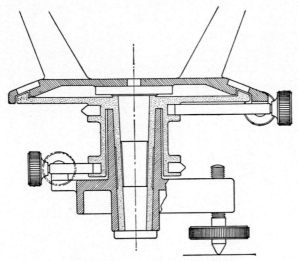

Fig. 4.9.—Part section of tribrach, etc.

point at which they are being read, is reflected into a microscope of
which the eyepiece lies near the eyepiece of the telescope, so that the
observer does not have to move from the end of the telescope to the
side of the instrument to take readings. Moreover, in some of the
larger instruments the system is designed to give directly the mean
of the readings at opposite ends of a diameter of the circle, thus replacing two separate readings by one. In addition, the graduations
can be etched cleaner and more sharply on glass than they can on silver,
and, as the light rays pass through the glass instead of being reflected
from its surface, a much finer image of the graduations, accompanied
by a much brighter field of view, is obtained. As a result, a higher
magnification can be used in the micrometer microscopes, and the
size of a circle to attain a given degree of accuracy can be considerably
reduced. Glass, however, has the disadvantage that in some parts

of the tropics it is subject to a fungus growth which eats into it and after a time makes the graduations illegible and the glass itself more or less opaque. Much research work has been done in recent years on *fungus*, and many makers now supply a special chemical which is kept in the instrument case and which is intended to prevent the growth of fungus on the surfaces of lenses and glass circles.

There are three main ways in which the graduations on the horizontal circles of theodolites are figured, and most makers offer a choice of system when supplying a new theodolite. In the first, the *whole-circle* system, the circle is figured to yield readings taken clockwise from 0° through 90°, 180°, and 270° to 360° (fig. 4.10*a*). In the second system, the *half-circle* system, the circle is figured right and left from 0° through 90°

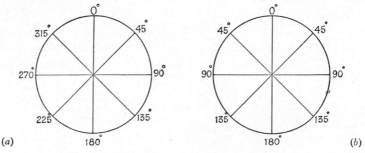

(*a*)   (*b*)

Fig. 4.10.—Examples of horizontal circle divisions

to 180° (fig. 4.10*b*). Occasionally a circle is figured on the *quadrantal* system in which the figures run in quadrants from 0° at each end of a diameter to 90° left and right of this diameter. Which of these systems is the best or most convenient depends mainly on the particular purpose for which the instrument is to be used. Ordinarily, the whole-circle system is the more convenient for all general survey work, but the second or third system is usually found to be better if the theodolite is to be used almost exclusively for work involving much laying out of angles, e.g. for the laying out of curves on roads and railways. The half-circle and quadrantal systems have the great disadvantage that, when angles are being measured and booked, it is necessary to note whether they are measured left or right from the first point sighted, and it is very easy to enter an angle as measured left when it is really measured right. With the whole-circle system there should be no possibility of such a mistake, since the angle is always to be taken clockwise from the first point sighted. Hence, the actual angle between two points is the reading on the circle when the second point is sighted less the reading when the first point is sighted, 360° being added to the reading to the second point if it is less than the reading to the first. The whole-circle system of figuring is the one most commonly used.

In theodolites fitted with micrometer microscopes, the figures above the graduations on the circle are inverted so as to appear upright when viewed in the eyepiece. In this case the figures in the image *appear* to run from left to right although they actually run from right to left on the circle itself.

12. The Upper Plate.

The upper plate consists of a circular metal disc mounted on a central spindle which fits into the hollowed-out portion of the spindle of the lower plate, so that the upper plate can be rotated relative to the lower plate about the centre of the spindle as axis. The upper plate carries two index marks, one at each end of a diameter, which are parts of two verniers or micrometers used for fine readings of the graduated arc on the lower circle. This plate also carries a level tube, or else two level tubes placed at right angles to one another, for levelling the instrument; two vertical standards, generally part of the same casting as the plate, for supporting the telescope and vertical circle; and sometimes, but not always, a compass.

The upper plate, complete with standards and telescope, is sometimes called the *alidade* of the instrument.

When verniers are fitted to a theodolite, the graduations on them are generally engraved on silver, and the silver strip comprising the vernier is inset into the outer rim of the upper plate so as to lie against, and in the same plane as, the graduated arc on the lower plate. The zero mark on the vernier is the index mark from which readings are measured. In nearly all cases, two verniers or micrometers are used set at opposite ends of a diameter of the circle so that at any setting the difference in reading by the verniers is almost exactly 180°. Each vernier or micrometer is generally marked with a distinctive letter, A or B, to distinguish it from the other.

The reasons for placing two verniers or micrometers at the opposite ends of a diameter are:

1. The readings on one may be used to act as a check against gross errors in readings on the other.
2. By taking the mean of the readings recorded by each vernier small errors due to eccentricity of the verniers or of the graduated arc with reference to the vertical axis of rotation, caused either by wear or by faults in construction, are eliminated.

The micrometer is an alternative method to the vernier for taking readings that involve splitting up the subdivisions on the horizontal circle, and a well-constructed micrometer is capable of giving much more accurate readings than a vernier. Moreover, a micrometer, as a general rule, is easier and quicker to use, but is bulkier, heavier, and more expensive: hence, verniers are usually fitted to the smaller and cheaper instruments, and micrometers to the larger and more accurate instruments. The micrometers on the upper plate are supported on small brass mountings screwed on to the plate or else attached to the standards, and, like verniers, are generally fitted in pairs, 180° apart.

The level tubes on the upper plate are similar to those mounted on an engineer's level (p. 121), but normally not so sensitive. One end of each tube is hinged and the other end bears a flat, horizontal lug with a hole in the centre (fig. 5.4b). A small threaded post, fixed at its bottom to the upper plate, passes loosely through this hole, and two small adjusting *capstan screws*, one above and one below the lug, enable the latter to be adjusted up or down and held firmly between the screws. The tops of the capstan screws consist of small cylinders of brass, with four radial holes bored in the outer rim. A thin metal rod, provided with the instrument and known as a *tommy bar*, fits into the holes and acts as a lever for turning the rings. Adjusting is done by loosening one ring and then tightening the other until the end of the tube is firmly held at the desired height on the post.

The clamp and fine-motion screw for clamping the lower and upper plates together, and for giving the upper plate a fine circular motion relative to the lower plate, are sometimes on top of the plate and sometimes at the side. In some instruments, however, they are below the lower plate, above the lower clamp.

Attached to the upper plate, or forming part of the same casting, are the two standards which support the telescope, vernier frame, and vertical circle. The telescope is mounted near its centre on a horizontal axis, called the *horizontal axis, trunnion axis,* or *transit axis,* this axis being at right angles to the main longitudinal axis of the telescope, with the ends resting in bearings on top of the standards. Each of these bearings consists of a block of hard metal, the upper surface of which is V-shaped, and the block rests in a rectangular recess in the top of each standard. The trunnion axis rests on the two sides of the V and is kept from sideways movement by the upright sides of the recess in the standard. On top of the standard is a metal clip, with a small pad of leather or cork fastened below it, which can be swung over and screwed down over the top of the axis so that the pad bears on the latter and keeps it in position in the bearing: in some cases the clip is replaced by a solid metal cap screwed down to the standard. One of the blocks on which the trunnion axis rests is adjustable up and down in the recess, and this motion provides the means of adjusting the trunnion axis to make it perpendicular to the vertical axis (see fig. 4.7, p. 66).

13. The Vertical Circle and Vernier Frame

Unless a theodolite is one of the modern types in which both the horizontal and vertical graduated circles are of glass, the vertical circle consists of a thin flat ring of brass in the outer surface of which is inserted a thin flat ring of silver on which the graduations are engraved. The brass ring is attached by arms or spokes to the horizontal axis carrying the telescope. Consequently, the graduated are rotates with the telescope when the latter is turned about the horizontal axis. An index, forming part of a vernier or micrometer, is attached to the standards through a *vernier frame* or *arm*, and remains fixed while the telescope and vertical circle are rotated about the horizontal axis, thus serving to indicate the reading on the circle. If the instrument is in adjustment, this reading should be zero when the theodolite is level and the line of sight truly horizontal. Hence, when this is so, the reading on the vertical circle, when the telescope is pointed to a station, is a measure of the vertical angle subtended at the instrument between the station and a horizontal plane through the horizontal axis.

As in the case of the horizontal reading system, the vertical reading system usually consists of two indices with verniers or micrometers set at opposite ends of a diameter. In the simplest form of construction, the vernier arm carrying the indices and verniers is fixed direct to one standard of the instrument, and reliance is placed on the level tubes on the horizontal plate for indicating when the instrument is level, no means of adjusting the set of the index and verniers being provided. More usually, a separate and specially sensitive level is provided on the vernier frame or on top of the telescope for use when vertical angles are being observed. In either case, a clamp and fine-motion tangent screw, enabling the vertical circle to be clamped to the vernier frame or to one of the standards, is fitted. When the vernier frame is not rigidly attached to one of the standards it is provided with an adjustable clip or antagonising screw which works against a lug on the standard and enables a fine-motion adjustment to be given to the frame and level relative to the standards. Sometimes the clamp and tangent screw for the vertical circle are fixed on the vernier frame,

4   (G 446)

so that the circle can be clamped to the frame and given a fine motion relative to it. A better balance, however, is obtained if the vernier frame is on the circle side of the telescope and the circle clamp and tangent screw on the other. In this last case (figs. 4.6 and 4.7, pp. 65–6), the clamping screw is at the top of a vertical arm, through which the horizontal axis passes, and the clamp serves to clamp the horizontal axis to the arm. The bottom part of the arm carries a tangent screw and antagonising spring which work against a lug fixed to the standard on that side of the telescope. Consequently, the telescope and vertical circle are held fixed relative to the standard when the clamping screw is tightened, but a fine motion relative to the standard can be given to them by means of the tangent screw.

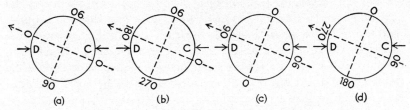

Fig. 4.11.—Examples of vertical circle divisions

The arrangement of the graduations on the vertical circle varies in different models, the principal ones being illustrated in fig. 4.11. In this diagram, the telescope is supposed to be pointing from right to left, with circle left, the angle of elevation being 20°. Remembering that the circle moves while the index and verniers remain fixed, it is easy to see how readings are to be taken. In fig. 4.11a, which represents a system usual in small vernier instruments, both verniers read 20°, the actual angle of elevation. In fig. 4.11b, C vernier reads 20° and D vernier 200°, so that 180° must be subtracted from the D reading to give the actual angle of elevation. If the angle had been one of depression, the readings would have been 340° on C and 160° on D. In figs. 4.11c and 4.11d, both circles are graduated to read zenith distances, or angles measured from the zenith. Hence, in fig. 4.11c, the readings for an angle of 20° elevation would be 70° on both verniers, and in fig. 4.11d they would be 70° on C vernier and 250° on D. It is also easy to see how the readings would run with circle right. Thus, with circle right and the telescope again directed from right to left, the readings in the system shown in fig. 4.11b would be 160° on C vernier and 340° on D for an angle of elevation of 20°. Here the circle, verniers and telescope are turned through an angle of 180° about the vertical axis and the telescope and circle then rotated back through 140° about the horizontal axis.

The system shown in fig. 4.11*a* is simple in use, and is convenient, since the reading on both verniers is the same, and is the same for both angles of elevation and depression. Consequently, when the instrument is used for determining vertical heights (but not when angles are only being observed in order to determine slope correction for chaining) the surveyor must be careful to note whether his observed angle is one of elevation or depression. In the system shown in fig. 4.11*b*, it is not necessary at the time of observation to note the sign of the angle, since this is at once apparent from the readings themselves. Thus, in addition to removing a possible source of ambiguity in reading and booking, this system offers a check against gross error, since the reading on the left-hand vernier is never the same as on the right-hand vernier. The systems illustrated in figs. 4.11*c* and 4.11*d* are commonly adopted in geodetic instruments where, in astronomical work especially, it is a little more convenient to work with zenith distances rather than with angles of elevation.

14. The Telescope.

The optical principles of the telescope depend ultimately on the fact that all parallel rays of light reaching a convex lens are bent when they leave it in such a manner that they intersect at a common point, the *focus,* and that all rays passing through another point, usually

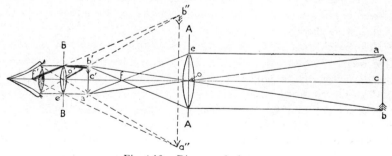

Fig. 4.12.—Diagram of telescope

situated near the geometrical centre of the lens, pass through the latter without bending. This latter point is called the *optical centre* of the lens, and the distance of the focus from the optical centre is called the *focal length* of the lens.

In fig. 4.12 AA is a convex lens whose optical centre is at **o** and focus is at **f**, and BB is another convex lens whose optical centre is at **o'** and focus is at **f'**. The focal length, **of**, is greater than the focal length **o'f'**. Let **a** be any point on a distant object and let **ae** be a ray from **a** parallel to **ofo'f'** striking the lens AA at **e**. This ray will be bent to pass through the focus **f** and will therefore travel from **e** in the direction **efa'**. Let **ao** be another ray from **a** passing through the optical centre **o** and meeting **efa'** in **a'**. Then, as **o** is the optical centre, the path of this ray will be a straight line, and hence **aoa'** is a straight line. The point **a'**, where the two rays **ea'** and **oa'** meet, will therefore be an image of the point **a**. Similarly, **b'** will be the image of a point **b**, another point on the object **ab**. Each intermediate point on this object will likewise form an image (approximately) on the line **a'b'**, and in this way **a'b'** becomes an upside-down (or reversed) image of the erect object **ab**.

Now consider the rays **a'e'** and **a'o'** which pass from **a'** parallel to **ofo'f'** and through the optical centre **o'** of the lens BB respectively. The ray **a'e'** which strikes the lens in a direction parallel to the line **ofo'f'** will pass through the focus **f'**. If the distance of the image **a'b'** from BB is less than **o'f'**, the focal length of BB, the rays **a'o'** and **e'f'** when produced backwards will intersect at **a''** on the right-hand side of BB. Hence, an eye placed near to BB between **o'** and **f'** will see an image of **a'** at **a''**. Similarly, an image of **b'** will be seen at **b''**, so that **a''b''** will be a magnified image of the image **a'b'**, and also of the object **ab**. Further, this image will be reversed with respect to **ab**.

The lens AA of the telescope which faces the object that is being viewed is known as the *object glass* or *objective*, and the lens BB, which is closest to the eye, as the *eye lens* or *eyepiece*.

The brightness of the image depends on the diameter of the objective relative to its focal length. For a given focal length, the larger the diameter of the objective the brighter the image.

The above description really applies to the ideal conditions where the lenses AA and BB are of negligible diameter and thickness. In practice, in order to obtain a bright image, it is essential that the diameter of the lens AA should be as large as possible, and it is impossible to make a lens which has no appreciable thickness. The result of this is that, if simple lenses were used, the telescope would have various optical defects, known as *aberrations*, which would result in curvature, distortion, unwanted colours, and indistinctness of the image. In order to eliminate these effects as much as possible, the objective and the eye lens are made up of two or more simple lenses. The front lens of the objective is a doubly convex lens made of crown glass, and the back lens is a concavo-convex lens made of flint glass, the two being cemented together at their common surface (fig. 4.13). The eyepiece in its most simple form usually consists of two plano-convex lenses, both of the same focal length, mounted in a tube with their curved surfaces facing one another, and separated by a distance equal to two-thirds the focal length of either lens (fig. 4.14). In front of the lens facing the eye, between it and the eye, is a small disc with a circular aperture slightly larger than the pupil of the eye.

Some theodolites and levels are fitted with special **erecting eye-pieces.** These give a magnified, but inverted, image of the image formed by the objective and hence, as the latter itself forms an in-verted image, the result is a magnified, but erect, image of the original

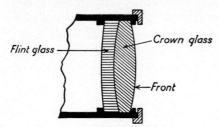

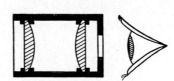

Fig. 4.13.—Section of object glass   Fig. 4.14.—Section of eyepiece

object. Such an eyepiece involves the use of extra lenses. This is a decided disadvantage since every extra surface between glass and air means the loss of a certain amount of light and hence a loss of brilliancy of the image. For this reason, erecting eyepieces are now seldom fitted to surveying instruments. Inverted images are not a great dis-advantage, and a surveyor very soon gets used to them and scarcely notices them.

The distance from the objective of the image formed by it is con-nected with the distance of the object by the relation

$$\frac{1}{v} + \frac{1}{u} = \frac{1}{f},$$

where

$v = \textbf{oc}'$ in fig. 4.12 (p. 79) $=$ distance of image from the optical centre,

$u = \textbf{oc}$ in fig. 4.12 $\qquad = $ distance of object from the optical centre,

$f = \textbf{of}$ in fig. 4.12 $\qquad = $ focal length of objective.

Hence, as f is a constant for the lens, the distance $\textbf{oc}'$ varies as the distance $\textbf{oc}$ of the object varies. It is essential in the construction of the surveying telescope that the image should always be formed in the fixed plane in the telescope where the cross hairs in the diaphragm are fitted. In order to accomplish this, the objective is made movable so that the point o can be moved closer to, or farther away from, c'. For this purpose, the objective is mounted in a sliding brass tube which moves freely inside and along a slightly larger tube. The cross hairs, or graticule, are fitted near the other end of the outer tube, and the eyepiece is fitted at the extreme end of this tube. A rack and pinion, operated by a milled head outside the outer tube, enables the objective to be moved in and out relative to the cross hairs until a clear image of the object is formed in the plane of the cross hairs. This operation is known as *focusing*.

The method of focusing just described, which involves a movement of the objective relative to the diaphragm, is known as *external focusing*. An alternative method, by means of which the objective is kept fixed, is now much favoured by manufacturers and surveyors. In this method a supplementary double concave lens is mounted in a short tube which can be moved to and fro between the diaphragm and the objective. This short tube holding the lens is moved along and inside the tube carrying the objective by means of a rack and pinion and an external milled head. This system has various advantages and disadvantages. The advantages are that an internal focusing telescope keeps out water and dust better than an external focusing telescope, and also lessens the effects of errors due to changes of position in the line of sight caused by a loose or worn objective or eyepiece mounting. The disadvantage is that there is the usual loss of brilliancy in the image caused by the interposition of the extra lens. On the whole, however, this disadvantage is outweighed by the advantages.

In older instruments, the *cross hairs*, which are designed to give a definite line of sight, consisted of short lengths of spider's web mounted on a flat metal ring which was suitably attached inside and to the telescope barrel. Spiders' webs, however, are easily broken and tend to sag when they get damp or old. Accordingly, as it is now possible to cut or etch very fine lines on glass, the spiders'-web diaphragm has been very largely replaced by a thin disc of glass, on which the cross hairs have been marked, mounted on a circular ring very similar to the ring used to mount spiders' webs (fig. 4.15). This ring is held in position in the barrel by four fine capstan-headed screws arranged at right angles to one another. The holes in the barrel through which the screws pass, which are made large in any case, are slightly elongated in the direction of the circumference so as to permit a certain amount of rotation when the screws are loosened. In this way, the position of the cross hairs inside the tube can be adjusted slightly, both horizontally and vertically, and a slight rotational movement is also possible.

Different arrangements of the cross hairs are employed. The

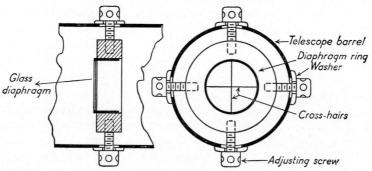

Fig. 4.15.—Diaphragm of telescope

simplest is a single central vertical line or hair with a single horizontal hair crossing it at or near the centre. This is shown to the left in fig. 4.16. Other arrangements are shown as (b), (c), (d) and (e) in the figure. The first three can be either web or glass, but the last two are only possible on glass. The upper and lower horizontal lines in (c), (d) and (e) are *stadia hairs* for determining distances, the use of these hairs being described later (pp. 149–151).

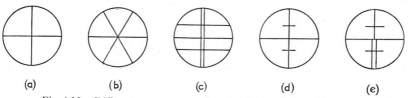

Fig. 4.16.—Different arrangements of cross hairs in telescope diaphragm

Broken cross hairs of spiders' webs can be replaced by means of a forked stick, a little shellac dissolved in spirit, and, last but not least, a live spider. First of all, clean the ring on which the new hairs are to be mounted, especially removing all old cementing material near the scratches which mark the positions for the hairs. Get the spider on the end of one of the forks of the stick and shake him off gently. As he falls, he will spin a thread, which can be fairly tightly wound round the prongs of the stick with a space between each part of the thread.

Immerse the threads for a few seconds in warm water and remove the superfluous moisture very carefully with a fine camel hair brush. Lay down the forked stick over the diaphragm ring, so that one length of thread lies exactly over the appropriate marks. Then fasten down by touching each end of the thread in turn with a drop of shellac and leave to dry. When the shellac is dry, cut the thread near the outer edges of the ring. The most difficult part of this operation often is to find and catch a suitable spider!

When a glass graticule is broken, it can only be replaced by a new one. Consequently, it is always well to carry one or two spare diaphragms or graticules in the case of the instrument, especially when working at considerable distances from where spare parts are obtainable. In many cases this means carrying complete spare diaphragm rings already fitted with etched glass graticules. In some of the instruments made by Messrs. Hilger and Watts, Ltd., however, a glass graticule is mounted in a thin metal socket which fits into the diaphragm ring. Spare sockets and graticules can be carried in special containers fitted in the instrument case, and a special extractor is supplied for removing the old graticule and putting in a new one.

The eyepiece must be adjustable so as to enable the cross hairs in the diaphragm to be brought into sharp focus. For this purpose, the eyepiece lenses, which have a much smaller diameter than the object glass, are contained in a metal tube which slides in and out along another slightly larger tube. Focusing with an ordinary sliding eyepiece is done by the fingers of one hand gently turning, and at the same time pushing or pulling the eyepiece in or out of its outer tube, no mechanical fine motion being fitted. More generally, however, the eyepiece is made to screw in and out of the main telescope barrel and is fitted at the outer end with a graduated ring somewhat larger than the diameter of the eyepiece tube. Focusing of the cross hairs is therefore done by turning the ring, and the position for exact focusing to suit any individual observer can be noted by means of the graduations on the ring and an index mark on the containing tube.

In order to avoid reflections in the inside of the telescope, and so to obtain as bright an image as possible, the internal surfaces of the various metal tubes used in its construction are painted a matt black.

It will be clear from the foregoing description of the telescope that *complete focusing* is done by first adjusting the eyepiece until the cross hairs are distinctly seen, and then the objective, or the supplementary lens in the case of the internal focusing instrument, is moved until a clear image of the cross hairs and object are seen in perfect focus together in the eyepiece. To test for perfect focus, move the head slightly from side to side and notice if the image of the object viewed moves with reference to the cross hairs. If there is any such movement, there is an error in focusing which is known as *parallax error*, and the objective must be refocused until the movement disappears. Thus, in fig. 4.17, suppose CC is the plane of the cross hairs and c one hair, and let the image formed by the objective lie in the plane AA, which is not coincident with the plane CC. Then with the eye at E, the cross hair will be seen against the point **a** on the image formed by the objective. When the eye is moved to E', the cross

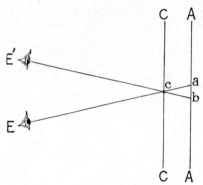

4.17.—Diagram illustrating parallax

hair will be seen against the object **b** and thus there is an apparent movement of the image formed by the objective across the hair, the movement being in the same direction as that of the eye when the objective image is between objective and cross hair, and in the opposite direction to the movement of the eye when the objective image lies between the cross hair and the eye. The apparent movement disappears when plane AA coincides with plane CC.

15. The Vernier

In fig. 4.18a we have two scales lying one against the other, 9 divisions of the long (lower) scale being of exactly the same length as 10 divisions of the short (upper) scale. Hence, if d is the length of a lower division and v is the length of an upper division,

$$9d = 10v$$

$$v = \tfrac{9}{10}d, \; d - v = \tfrac{1}{10}d.$$

Consequently, if we call the short scale the *vernier scale* and the long scale the *main scale*, the first division beyond zero on the vernier scale falls short of the first division on the main scale by $\frac{1}{10}$ of one division of the latter. The second mark on the vernier scale falls short of the second mark on the main scale by $\frac{2}{10}$ of a main scale division and so on. From this it will be seen that, if the vernier scale is moved to the left as shown in fig. 4.18*b*, so that, for instance, the 6th mark on it coincides with the 6th mark on the main scale, the index mark

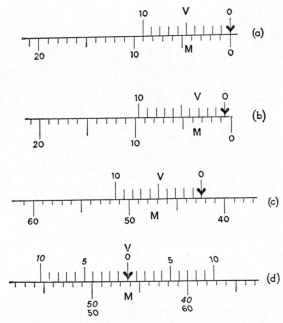

Fig. 4.18.—Diagrams to illustrate the theory of the vernier

of the vernier scale will have moved to the left by $\frac{6}{10}$ of a division of the main scale. In this case, the reading of the index mark will be 0·6, where the unit of length is the length of one of the small main scale divisions. Similarly, in fig. 4.18*c*, the index on the vernier is between the 42nd and 43rd graduations on the main scale, and the 4th graduation on the vernier scale is in coincidence with the graduation 46 on the main scale, which is 4 graduations on from the 42nd graduation. Hence, the reading is 42·4.

We can generalize this as follows:

Let n *of the smallest divisions on the vernier scale be equal in length* to (n — 1) *of the smallest divisions on the main scale, and let* d *be the value (inches, tenths, minutes of arc, etc.) of one division on the main scale. Let the index on the vernier scale come somewhere between the* P*th and* (P + 1)*th graduation on the main scale. Look at the vernier scale and find which one of its graduations coincides with a graduation on the main scale. Let this be the* r*th graduation on the vernier scale. Then the vernier reading is* $P + r\dfrac{d}{n}$.

The fraction $\dfrac{d}{n}$, sometimes called the *least count*, is the measure of the smallest unit which the vernier will resolve. To obtain the least count, *find the value of the smallest division on the main scale and divide it by the number of divisions on the vernier.* This is a constant for the vernier so that the reading on the latter is always r times the least count plus the lower of the two readings on the main scale between which the index lies.

Fig. 4.18*d* shows a *double vernier* in which the left-hand vernier (reading 46·3) is used in conjunction with the upper figures on the main scale (those sloping to the right) and the right-hand vernier (reading 53·7) in conjunction with the lower figures on the scale (those sloping to the left).

Sometimes the vernier divisions are longer, instead of shorter, than the divisions on the main scale. Thus, let 10 divisions on the vernier scale be the same length as 11 divisions on the main scale. Then,

$$11d = 10v,$$

$$v = \tfrac{11}{10}d, \; v - d = \tfrac{1}{10}d.$$

In this case, the graduations on the vernier run in the opposite direction to the graduations on the main scale as shown in fig. 4.19*a*. In fig. 4.19*b* the graduations on the two scales coincide at the graduation 4 on the vernier, and the index lies between 41 and 42. Hence, the reading is 41·4. This type of vernier is called a *retrograde vernier*, and the rule for reading it is the same as before. Note that this vernier is not to be confused with the right-hand vernier in fig. 4.18*d*, which

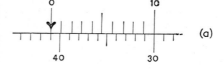

Fig. 4.19.—Diagram of retrograde vernier

is a direct vernier, as it reads to the right, but for a scale also increasing to the right.

It may happen that the divisions on the main scale are very close, and it would then be difficult, if the vernier were of normal length, to judge the exact graduation where coincidence occurred. In this case, an *extended vernier* may be used. Here $(2n - 1)$ divisions on the main scale are equal to n divisions on the vernier, so that

$$nv = (2n - 1)d,$$

$$v = \frac{(2n - 1)}{n} d, \ 2d - v = \frac{d}{n}.$$

This is the same value for the least count as that already obtained for a vernier of normal length, and hence, having found the least

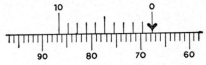

Fig. 4.20.—Diagram of extended vernier

count, the vernier can be read exactly as before. Fig. 4.20 shows such a vernier. The 10 divisions on the vernier scale are the same length as 19 divisions on the main scale. Each division on the main scale represents 1 unit, and there are 10 divisions in the vernier. Hence the least count is $\frac{1}{10}$, and, as the seventh graduation on the vernier scale coincides with a graduation on the main scale, and the index is between 67 and 68, the reading is 67·7.

Fig. 4.21 shows a vernier belonging to the horizontal circle of a theodolite. The smallest division on the main scale is half a degree,

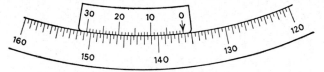

Fig. 4.21.—Diagram of vernier on graduated arc

or 30′, and there are 30 divisions on the vernier corresponding to 29 divisions on the main scale. The least count is therefore $\frac{30}{30} = 1'$. The index lies between 136° and 136° 30′ and the graduation 12 on the vernier coincides with a graduation on the main scale. Hence the reading is 136° 12′. Had the index come between the graduations

136° 30′ and 137°, the reading would have been 136° 30′ + 12′ = 136° 42′.

The student is recommended to draw on a large scale, say using a circle of about 8 in. radius, or even if necessary a straight line, one or more of the following vernier arrangements, which are often used on theodolites:

d	n	Least Count
30′	60	30″
20′	60	20″
10′	60	10″ (Extended Vernier).

Small magnifiers are usually supplied with the ordinary vernier theodolite, one to each vernier, to enable the verniers to be read easily. These are attached to special arms which permit a certain amount of lateral motion along the vernier scale, so enabling the magnifier to be brought exactly over any desired graduation.

When using a vernier, care must be taken to avoid looking at the graduations sideways, as otherwise a false estimate of the point of coincidence may be formed. If it is still difficult to decide the exact graduations which are in coincidence, an examination of the positions of the first two or three graduations on either side of the supposed point of coincidence, relative to the adjoining graduations on the main scale, will often enable a good estimate to be made. As these graduations get farther away from the point of coincidence, the marks on the one scale gradually separate from those on the other. The point about which equal widths of the small spaces between the graduations on the two scales are disposed symmetrically on either side is the point required.

The vernier scale itself cannot be adjusted, so that it is important that the length of the vernier scale should be correct, and that the intermediate graduations should be evenly and accurately spaced.

16. The Micrometer Microscope.

As a general rule, verniers are fitted to theodolites when the finest reading to be taken is not less than 20″, although vernier instruments are occasionally to be found which read to 10″. When, however, the finest reading is to be less than 20″, the micrometer microscope ordinarily replaces the vernier. In this way, readings may be taken on large geodetic theodolites to 1″ and estimated to 0″·2 or 0″·1, and on smaller theodolites the readings may be taken direct to 10″ or 5″ and estimated to 2″ or 1″.

The micrometer microscope (figs. 4.22 and 4.23) consists of a small low-powered microscope fitted with a small rectangular metal box at a point near where the image of the graduations formed by the objective will be situated. This box, which has rectangular or circular openings or windows in the top and bottom where it comes inside the tube of the microscope, is fitted with a fixed mark or index and with a movable slide carrying a vertical hair, or pair of parallel hairs placed very close together. These hairs are fitted so that they will lie parallel to the images of the division marks of the graduated arc. The slide is moved by means of a screw operated by a milled head and gradu-

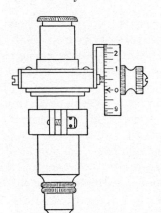

ated drum on the outside of the box, the pitch of the screw being such that a complete revolution moves the slide through a space equal to that between the images of two successive divisions of the gradu- ated arc when these images are formed in the plane containing the index mark

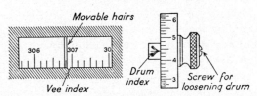

Fig. 4.22.—Micrometer microscope

Fig. 4.23.—Reading micrometer microscope

and movable hairs. Springs on the side of the slide opposite the screw and working against the former are fitted to take up back- lash. Fractional parts of a revolution of the drum, corresponding to fractional parts of a division on the horizontal circle, may be read on the graduated drum against an index mark fitted to the side of the box. In some micrometer microscopes, the eyepiece is directly above the box, the whole microscope being inclined at a small angle away from the vertical, but in others, for the sake of convenience in reading, a prism placed above the box serves to reflect the image of the hairs and divisions through an eyepiece placed at an angle to the horizontal. The function of the eyepiece is, of course, to form a magnified image of the index, movable hairs and the image of the graduations formed by the obejctive.

In order to provide a means of focusing and adjusting for *run*, the microscope is mounted in such a way that it can be moved up and down in its support and held in any position. Thus, in fig. 4.22, the capstan-headed screw in front may be used to loosen or tighten the slotted ring holding the microscope to its support. In addition, the objective may be moved up and down in the tube which carries it by means of the knurled objective mounting and clamping rings shown in fig. 4.22. These two movements permit the dimensions of the image of the graduations and of the spaces between them to be altered slightly and at the same time the image to be brought into focus on the plane containing the cross hairs. As a result, it will be seen that, although the screw working the slide must be evenly cut so that the thread is uniform throughout, it is not essential that its pitch, or distance between successive threads, should be exactly its nominal value.

The method of using and reading the micrometer will be understood from fig. 4.23, which shows the view in the eyepiece, together with the graduations on the adjoining drum. Some of the graduations on the horizontal circle are seen through the rectangular window, and near the centre of the bottom of the window is a small V notch which serves as an index. When the drum reads zero, one of the graduations should be in the centre of the V, and this graduation should appear to be central between the two movable hairs. One complete revolution of the drum moves the cross hairs from one graduation to the next one.

In this example, the circle is graduated to 10′ of arc, the graduated drum is divided into 10 large intervals, and each of the large intervals into 6 small ones. Consequently, each of the large divisions on the drum corresponds to 1′ of arc and each of the small divisions to 10″ of arc. To read the instrument, note the divisions on either side of the V and take the lower one. Then turn the drum until the nearest division seems to be midway between the two vertical hairs and note the minutes and seconds indicated against the index at the side of the graduated drum, estimating single seconds by eye. Thus, in the example, the V comes between 306° 50′ and 307°, and is a little nearer the graduation 306° 50′. Turn the drum until the graduation 306° 50′ lies midway between the movable hairs. The index beside the drum is now between the graduations 4′ 20″ and 4′ 30″, and, by estimating tenths, the reading on the drum may be taken as 4′ 26″. Hence the complete reading is 306° 54′ 26″.

Other different arrangements of graduations on the circle and drum are often employed, but the general principles are the same as those just described, so that the student should have no difficulty in dealing with any other case that he may meet with.

There are four adjustments necessary in the micrometer microscope, these being:

1. To focus.
2. To make the vertical hairs parallel with the image of the graduations of the circle.
3. To make the drum read zero when a graduation is in the centre of the V index.
4. To adjust the image of the graduations of the circle so that the space between two successive graduations corresponds exactly with the space traversed by the cross hairs during one complete revolution of the drum. This adjustment is generally known as the *adjustment for run.*

In *focusing*, the first step is to focus the movable hairs in the eyepiece. This is done by drawing the eyepiece out of the tube in which it slides, or pushing it in, until a clear, sharp image of the vertical movable hairs is obtained. A piece of white paper slipped under the microscope immediately on top of the graduated arc will enable the hairs to be seen more clearly. In some instruments, the eyepiece is screwed into the outer tube, and focusing is done by rotating the whole eyepiece mounting. When what seems to be a sharp image of the hairs has been obtained, close the eye for a few seconds and then look again to see if the image is still sharp.

Having focused the eyepiece, move the whole microscope up or down in its mounting by loosening the capstan-headed screw provided for the purpose until a clear image of the graduations is seen, and this image does not move relative to the hairs as the eye is moved slightly from side to side. Then tighten the capstan-headed screw. This completes the focusing of the microscope.

Now try bringing the cross hairs over the image of one of the long degree graduations of the circle. This graduation should be *exactly parallel* to the hairs. If it is not, loosen the microscope slightly in its support by means of the capstan adjusting screw in the supporting split ring, and turn it until the graduations and the hairs appear to be parallel. Tighten the screw when this position has been secured.

To adjust the *zero of the drum*, clamp both of the horizontal plates and use the tangent screw of the upper plate to bring one graduation in the exact centre of the V index. Bring the cross hairs over this graduation and note if the reading on the drum is zero. If not, loosen the screw outside the milled head of the drum, so permitting the latter to be revolved without moving the hairs. Turn the drum until the zero mark is opposite the index, and, the graduation still appearing to be central between the hairs and in the centre of the V index, tighten up the screw of the milled head again.

The last adjustment (*adjustment for run*) is to ensure that the interval between the images of two successive graduations of the circle is the same as the distance traversed by the hairs when the graduated drum is moved from zero to zero through one complete revolution. Set the drum at zero with the hairs evenly astride the image of a circle division falling in the V index. Revolve the drum until the hairs lie evenly on either side of the next division on the circle. Observe the reading on the drum, and if this is not zero, the micrometer requires adjustment. If the reading is too large, the objective is too far away from the graduations and needs to be lowered towards them: if the reading is too small, the objective should be moved farther away from the graduations. By means of the knurled clamping and adjusting rings move the objective nearer to or farther away from the graduated arc. This will alter the dimensions of the image of the graduations and will upset the adjustment for focusing. Bring the graduations into proper focus by moving the whole microscope in its support until the image of a graduation does not move relative to the cross hairs as the head is moved slightly from side to side. Repeat the test, and, if it is not satisfied, repeat the adjustment in the same manner as before.

Most micrometer microscopes will retain the above adjustments for a considerable time, with the exception of the adjustment of the eyepiece for focusing the cross hairs and the image of the graduations. Consequently, focusing the eyepiece is the only adjustment which may be necessary at individual set-ups, or at the beginning of a day's work.

17. Vernier Theodolite by Messrs. W. F. Stanley.

The theodolite shown in fig. 4.7 (p. 66) is a simple 5-in. vernier theodolite or tacheometer by Messrs. W. F. Stanley. Both circles are 5 in. diameter with verniers reading to 20″ of arc. The horizontal circle is completely enclosed in a dust- and water-tight cover, the verniers and graduations on the arc being visible through small removable glass windows in the upper plate cover. Magnifiers for each of the four verniers are provided. The three main levelling screws also are provided with dust caps to keep out dust and wet, and provision is made for adjusting tightness and taking up wear. The centring device is between the trivet stage and tribrach, and gives a range of movement of about 1 in. Two level tubes are fitted, one on the horizontal plate and one on the vernier arm, the latter having a sensitivity of

20″ per $\frac{1}{16}$ in. The internal focusing telescope has an equivalent focal length of 10 in., an objective of 1·65 in. diameter, an angular field of 1° 45′, and a magnification of about 24 diameters. Sights for rough pointing and a mark for overhead plumbing are fitted on the telescope. The adjusting screws of the diaphragm are not visible in this instrument as they are protected by a detachable ring cover seen between the rear sight and the eyepiece on the telescope. This eyepiece is screw-focusing and carries a scale for quick focusing. Various extras, such as a trough compass, a quick-setting levelling head instead of the ordinary levelling head, and a diagonal eyepiece, can be had at extra cost.

18. Micrometer Theodolite by Hilger and Watts.

Fig. 4.24 (Plate VI) shows a small micrometer theodolite or tacheometer manufactured by Messrs. Hilger and Watts, Ltd. In this theodolite, the horizontal circle is 6 in. diameter, graduated to read 5″ of arc per division on the micrometer drum, and by estimation to 1″, and the vertical circle, which is totally enclosed, is 5 in. diameter, also graduated to read to 5″. The telescope, which is internally focusing, has an objective of 1·6 in. diameter and a magnification of 27 diameters. A folding mirror over the upper level tube enables the bubble to be seen from the eyepiece end of the telescope, this bubble having a sensitivity of 8″ to 10″ per 2 mm., and the bubble on the horizontal plate a sensitivity of 20″ per 2 mm. The centring device on this instrument has a range of movement of 2 in., which is rather an unusually large amount for an instrument of its kind.

It will be noticed that, for convenience in reading, the eyepieces of the micrometer microscopes on the horizontal circle are at an angle with the main part of the microscope barrel, the light rays being reflected at the bend by a reflecting prism. Inside the small glass tubes which can be seen between the upper circle cover and the objective of each micrometer is a small tube cut away at an angle. The inside of this tube, which is rotatable about its vertical axis, has a white reflecting surface, so that it reflects light from outside on to the circle graduations, and thus makes them visible in the eyepiece of the micrometer.

19. Theodolites with Glass Circles and Optical Micrometers.

The advantages of glass circles instead of the usual silvered graduated arcs have already been referred to (p. 73), and several makers now provide a variety of models equipped with such circles. In nearly all these models, the eyepiece of the microscope or micrometer, if

PLATE VI

Fig. 4.24.—MICROMETER THEODOLITE BY MESSRS. HILGER
AND WATTS, LTD.

(By courtesy of the makers)

not alongside the telescope, can be swung round to be parallel to it, so that after the surveyor has sighted through the telescope he does not have to change his position to one side to read the micrometers. In addition, the larger instruments, such as the Cooke, Troughton & Simms " Tavistock " and the Zeiss and Wild theodolites, are so designed that the optical micrometer, or optical reading device, gives directly a reading which in effect is really the mean of readings at opposite ends of a diameter, thus eliminating any error which might otherwise arise from non-coincidence of the centre of the graduations and the axis of rotation of the instrument. In the smaller instruments, however, the reading is from one end of a diameter only, but in this case special precautions in design and construction are taken by the makers to ensure that the centre of the graduations and the axis of rotation are in coincidence and that the setting is not likely to be upset by temperature changes.

When a theodolite is fitted with a glass circle, rays of light generally pass through the glass of the circle, and are thence reflected by prisms or mirrors to the micrometer. Hence, means must be provided for reflecting ordinary sunlight into the interior and then through the circle and micrometers. Also, electric illumination provided by dry batteries or accumulators is usually fitted to supplement natural illumination.

Messrs. Cooke, Troughton & Simms make several instruments with glass circles, the earliest and best known of these being the Tavistock theodolite, of which there are two models—the geodetic model and the ordinary model. The horizontal and vertical circles in the smaller model are divided to 20′ and are of $3\frac{1}{2}$ and $2\frac{3}{4}$ in. diameter respectively. In the micrometer of this instrument, graduations from opposite ends of a diameter are brought into the same field of view so that they appear to be close together on either side of a fine index line, readings being taken direct to 1″. Separate micrometers for the horizontal and vertical circles are employed, and the eyepieces of these are pivoted so that the observer can view them from his position near the eyepiece of the telescope whether the latter is in the direct or reverse position.

The Cooke, Troughton & Simms Optical Micrometer Theodolite No. III, shown in fig. 4.25 (Plate VII, p. 96), is a somewhat smaller instrument than the Tavistock, and enables angles to be observed direct to 20″ and by estimation to 5″, the reading in this case being taken from only one end of a diameter. Fig. 4.26 shows the appearance in the field of view of the micrometer, which is common to both horizontal and vertical circles. Three windows are seen, the larger one on the upper right, marked V, being for use with the vertical circle, and the large one on the lower right, marked H, being for use with the horizontal circle. Turning a milled head situated on one standard below the altitude spirit level enables the image of a main division to be moved and brought into contact with the index in the large window. This gives the degrees, and the reading of the index in the small window on the left gives the minutes and seconds. Thus, in fig. 4.26, the reading on the horizontal circle is 312° 29′ 40″.

The C.T. & S. Theodolite No. III can be supplied in two forms. One is the ordinary form with the upper part of the instrument permanently attached to the levelling head. In the other form, the upper part of the instrument, including lower and upper circles, is detachable from the levelling head. This is to enable it to be used with the *three-tripod system* of observing. In this system, three or more tripods are used, each with a separate levelling head. Special observing targets for sighting on can be mounted on the levelling heads, and they and the upper part of the theodolite are interchangeable on these heads. This method of observing angles is more rapid than the ordinary method provided sufficient

Fig. 4.26.—Micrometer readings in Cooke, Troughton and Simms No. III theodolite.
(By courtesy of the makers)

labour is available, and it has the advantage that errors due to faulty centring over a ground mark are reduced to a minimum, a point of some importance when, as in mining work, sights are sometimes very short.

Messrs. Hilger and Watts, Ltd., manufacture three models with optical micrometers, these being known as the " Microptic " Theodolite, Nos. 1, 2 and 50. No. 2 is the largest and most accurate of the three instruments as it enables readings direct to 1″ to be taken at opposite ends of a diameter, whereas the readings on the other two are taken at one end of a diameter only, direct to 20″, and by estimation to 5″ in the case of No. 1, and direct to 1′ and by estimation to 0′·1 (6″) in the case of No. 50.

Fig. 4.27 (Plate VIII, p. 98) shows the Watts Microptic Theodolite No. 1, the view in the micrometer being shown in fig. 4.28. Here there

PLATE VII

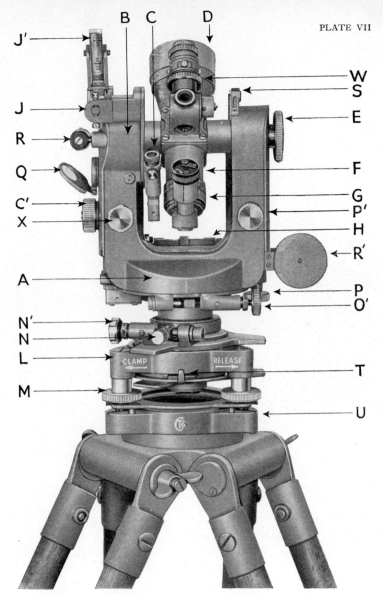

Fig. 4.25.—COOKE, TROUGHTON AND SIMMS OPTICAL
MICROMETER THEODOLITE No. III (approx. ⅓ full size)

A. Casing of horizontal circle.
B. Casing of vertical circle.
C. Circle reading eyepiece.
C′. Micrometer milled head.
D. Object glass.
E. Focusing head for telescope.
F. Screw focusing eyepiece.
G. Cover to reticule adjusting screw.
H. Plate spirit level.
J. Altitude spirit level.
J′. Reversible mirror to J.
L. Tribrach.
M. Footscrew.

N. Clamp for lower plate.
N′. Slow-motion screw for lower plate.
O′. Slow-motion screw for upper plate.
P. Clamp for telescope.
Q. Illumination reflector for circles.
R. Electric lamp for illumination of circles
    and altitude level.
R′. Battery for electric illumination.
S. Facing for magnetic compass.
T. Clamp for centring motion.
U. Trivet stage.
W. Optical plummet.

(By courtesy of the makers)

are three windows, the upper one giving an image of the vertical circle, the middle one an image of the horizontal circle, and the bottom one giving the minutes and seconds direct to 20″. The circles are divided into intervals of 20′, and a milled head is turned until the nearest 20′ division is brought accurately to the middle of the space between the two fine index lines. The extra minutes and seconds are then read against an index line in the lower window. Readings can easily be estimated to 10″, and after a little practice to 5″. In the illustration, the reading of the horizontal circle is 23° 32′ 30″.

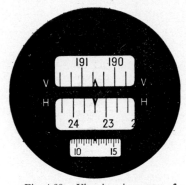

Fig. 4.28.—View in micrometer of Watts Microptic Theodolite No. 1.

(By courtesy of Messrs. Hilger and Watts, Ltd.)

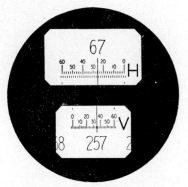

Fig. 4.29. — View in micrometer of Cooke, Troughton and Simms Theodolite No. 4.

(By courtesy of the makers)

20. Theodolites with Estimating Microscopes or Optical Scales.

The estimating microscope or optical scale is simpler in construction than the optical micrometer and does not give such precise readings. In general, however, it is slightly more accurate than the vernier, but, compared with the latter, it tends to add to the weight and bulk of the instrument, besides being more expensive. It consists of a small microscope in which a glass diaphragm, on which a special scale has been engraved, is placed in the usual position in the focal plane of the eyepiece. An image of the graduations is formed in this plane, and the odd minutes between the nearest division on the circle and the zero of the scale are read off direct on the latter.

The Optical Scale Theodolite No. IV of Messrs. Cooke, Troughton & Simms is provided with a horizontal glass circle of 3 in. diameter and a vertical glass circle of 2 in. diameter. Each circle is graduated to single degrees, the minutes and seconds being read on the scales in the estimating microscope. These scales are seen in two rectangular windows (fig. 4.29) in the field of view of the microscope, the readings on the horizontal circle being given in the upper window and those of the vertical circle in the lower one. There are two parallel scales used for obtaining the reading of the horizontal circle. The upper scale has 30 divisions and gives the even minutes, and the lower scale, which is off-set relative to the upper by an amount equivalent to one minute and has 31 divisions, gives the odd minutes. When the circle graduation mark coincides exactly with an upper or lower mark on the scale, the minutes can be read directly, but when the circle graduation appears to lie midway between a graduation on the upper scale and one on the lower scale, the reading is taken as so many degrees and minutes plus thirty seconds. Thus, in the example, the reading on the horizontal circle is 67° 25′ 30″.

The scale for the vertical circle is single, and is divided into 30 divisions so that the value of one division is 2′. Hence, readings may be taken direct to 2′ and by estimation to 1′. With a little practice, however, the horizontal circle reading may be estimated to 15″ and the vertical to 30″.

21. The Wye Theodolite.

All the theodolites we have described above are of the transit type. That is to say, the telescope can be *transited* or turned through a complete revolution, or through almost a complete revolution,* about a horizontal axis which is attached to, and can be considered to be part of, the telescope. In the instrument known as the wye theodolite, however, the horizontal, or trunnion, axis is not part of the telescope, and the latter is carried on two U- or Y-shaped cradles, similar to the cradles supporting the telescope of a wye level (p. 129), from which it can be removed and turned end for end or about its longitudinal axis. These cradles are supported on a plate above the vertical circle, which in this case consists of a semicircle mounted on a horizontal axis. Otherwise, the construction of the wye theodolite is very similar to that of the ordinary transit theodolite.

Wye theodolites are very seldom used nowadays except by a few old-fashioned firms of engineers or surveyors who happen to have a theodolite of this type in their possession. Consequently, we do not propose to describe the wye theodolite in detail. Its adjustments are somewhat different from those of the transit theodolite, and its

---

* In some transit instruments the object-glass end of the telescope prevents complete transiting, as the upper part of the upper plate and its attachments come in the way when this end of the telescope is depressed. In such a case, the telescope can be transited by depressing the eyepiece end.

PLATE VIII

Fig. 4.27.—WATTS MICROPTIC THEODOLITE No. 1

1. Bubble mirror.	6. Telescope clamp.	10. Plate bubble.	15. Lower clamp.
2. Alidade bubble.	7. Circle eyepiece.	11. Upper clamp.	16. Daylight reflecting
3. Bubble reflector.	8. Optical micrometer	12. Upper plate tangent	mirror.
4. Pointer sights.	milled head.	screw.	17. Alidade adjustment.
5. Telescope focusing	9. Telescope fine ad-	13. Levelling screws.	18. Eyepiece of optical
ring.	justment.	14. Lower plate tangent	micrometer.
		screw.	

(By courtesy of Messrs. Hilger and Watts, Ltd.)

only advantage appears to be that the adjustment for collimation is a little simpler than the corresponding one for the transit theodolite.

22. Adjustments of the Theodolite.

Most theodolites are inclined to get out of adjustment from time to time, and accordingly the surveyor must always be prepared to undertake the following tests and " permanent " adjustments as and when they are required or thought necessary:

1. *Adjustment of Horizontal Plate Levels.*—To make the horizontal plate levels perpendicular to the vertical axis.
2. *Collimation Adjustment.*—To make the line of sight through the cross hairs and optical centre of the objective perpendicular to the horizontal axis.
3. *Horizontal Axis Adjustment.*—To make the horizontal axis of the telescope perpendicular to the vertical axis.
4. *Adjustment of Telescope Level.*—To adjust the level on the telescope so that it is parallel to the line of sight or collimation.
5. *Vertical Circle Index Adjustment.*—To adjust the index on the vernier frame so that the circle reads zero when the line of collimation is horizontal.

In addition, the following " temporary " adjustments must be made every time the instrument is set up:

1. *Centring.*—To centre the vertical axis accurately over a mark on the ground or under a mark overhead.
2. *Levelling.*—To level the instrument so that the vertical axis is truly vertical, and to make the level on the vernier frame, if one is fitted and vertical angles are to be observed, perpendicular to the vertical axis.
3. *Focusing.*—To focus the eyepiece and objective of the telescope so that a clear image of the cross hairs and of the object is obtained free from parallax (p. 85).

Permanent Adjustments

Adjustment of Horizontal Plate Levels.—This adjustment involves making the vertical axis vertical and then making the levels on the upper plate perpendicular to the vertical axis. When this adjustment has been completed, the bubbles of the plate levels will remain in the centre of their run as the instrument is turned about the vertical axis through any angle.

Having set up the instrument as level as can be judged by eye and loosened the clamps of both horizontal circles, turn it about the vertical axis until the longer level on the upper plate is parallel to the plane containing the axes of one pair of levelling footscrews Bring the bubble of the level to the centre of its run by means of these two screws, turning one in a clockwise direction and the other in an anticlockwise direction, but turning both together. Rotate the instrument about the vertical axis until the level is at right angles to its original position, and bring the bubble to the centre of its run by means of the third footscrew alone, or by the other pair of footscrews if the instrument is one with four screws. This will make the plate approximately level. Now turn the instrument back to its original position, and, if necessary, bring the bubble to the centre of its run by means of the original pair of footscrews. Then turn the instrument about the vertical axis through 180°, so that the level tube is parallel to its original position but with the ends pointing in the opposite direction to the direction they were in originally. If the bubble is no longer in the centre of its run, half of the error is due to the vertical axis not being vertical and half to the level not being perpendicular to the vertical axis. Hence, correct half the error by means of the footscrews and then make the bubble central by means of the capstan screws which are fitted at one end of the tube for this adjustment.

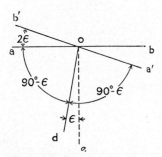

Fig. 4.30.—Adjustment of bubble and vertical axis

The reason for the above procedure can easily be seen as follows:

Suppose the bubble is not perpendicular to the vertical axis and that the end **oa** (fig. 4.30) makes angle $90° - \epsilon$ with the vertical axis **od**. In the first position of the level, the latter, after levelling, will be represented by the horizontal line **ab**, and **od** will be the position of the vertical axis of rotation, where angle **aod** $= 90° - \epsilon$. The vertical axis will therefore make an angle ϵ with the true vertical **oe**. When the instrument is turned through 180°, the rotation will be about **od** so that the end **a** will come to **a′**, where **a′od** $=$ **aod** $= 90° - \epsilon$, and the bubble will occupy the position **a′ob′**. Hence **boa′** is the amount by which the bubble is now off its centre. But **boa′** $= 180° -$ **aoa′** $= 180° -$ **aod** $-$ **a′od** $= 180° - (90° - \epsilon) - (90° - \epsilon) = 2\epsilon$. If, therefore, half of this error is corrected by the footscrews, the vertical axis **od** will be brought to **oe** and so made vertical, and, by correcting the other half of the error by means of the capstan screws on the level tube, the bubble will be brought to a truly horizontal position at right angles to the vertical axis.

Having completed the adjustment over one pair of screws, the instrument should be turned through 90°, so that the level is parallel to the vertical plane that contains the vertical axis and the axis of the third screw, and is perpendicular to the vertical plane containing the axes of the other two, and the bubble brought to the centre of its run by means of the third footscrew. The axis should then be truly vertical in all vertical planes passing through it, so that the bubble should remain in the centre of its run in all positions of the instrument as the latter is revolved through 360° about the vertical axis. If it does not do this, the test and adjustment should be repeated over another pair of footscrews. After the longer level tube has been adjusted and the vertical axis made vertical, the second level, if one is fitted, can be adjusted by bringing the bubble to the centre of its run by means of the adjusting screws at the end of the tube.

Collimation Adjustment.—Before proceeding with this adjustment it is advisable to see if the vertical and horizontal hairs are truly vertical and horizontal when the instrument is levelled up. For this purpose, suspend a plumb bob with a long string attached to it a short distance away. Level the instrument carefully, if necessary using the sensitive level on the telescope or vernier frame for the purpose (p. 109), and point the telescope towards the string of the plumb bob. Carefully focus the telescope so as to obtain a clear, sharp image of the string free from parallax (p. 85) and note if this image is parallel to the vertical hair of the diaphragm. If not, loosen the capstan screws of the diaphragm (fig. 4.15) and turn the latter until the vertical hair and the image of the string are parallel or coincident throughout the length of the hair. Then tighten up the screws of the diaphragm, at the same time insuring that the vertical hair remains vertical while this is being done.

To test the horizontal wire, see that the instrument is properly level, and point it so that the horizontal hair intersects a mark, such as a nail head, some distance away and at about the same elevation as the telescope. Then note if the hair remains on the image of this mark as the instrument is slowly rotated about the vertical axis. If it does not do so, the horizontal hair is not perpendicular to the vertical hair. Nothing can be done about this except to put in new hairs, but this will be unnecessary if all observations for vertical angles are taken to the point of intersection of the hairs, not to any random point on the horizontal hair.

In order to carry out the adjustment proper, set up the instrument on flat ground and level up carefully. Drive a nail in a peg about 400 ft. away, A in fig. 4.31a, and turn the instrument to sight A with circle left, that is, with the vertical circle on the left-hand side of the observer. Clamp the lower and upper horizontal plates, and, after careful focusing, bring the vertical cross hair into coincidence with A by means of the fine-motion tangent screw of the lower plate. Then transit the telescope (i.e. turn it over through approximately 180° about its horizontal axis) until it points in the opposite direction. Suppose the line of sight **o**A is not perpendicular to the horizontal axis **aob** but makes an angle A**o**a = 90° − e with it, so

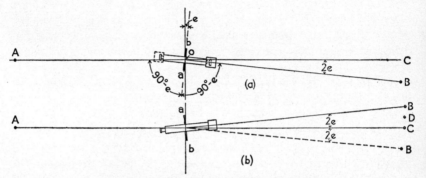

Fig. 4.31.—Collimation adjustment of theodolite

that **aob** makes an angle e with a line drawn perpendicular to **o**A through **o**. As the rotation takes place about the axis **boa**, the line of sight when the instrument is transited will come to a position **o**B such that B**oa** = A**oa** = 90° − e. Drive a peg at some convenient point B, such that **o**B is approximately equal to **o**A, and put a nail in it so that the image of the nail is intersected by the vertical hair. Then the angle e is the collimation error, and, if A**o** is produced to C, angle C**o**B = 180° − A**o**B = 180° − A**oa** − **ao**B = 180° − (90° − e) − (90° − e) = 2e. Loosen the lower clamp and swing the instrument about the vertical axis so that it again sights A (fig. 4.31b), the instrument now being circle right, that is, with the vertical circle on the right-hand side of the observer. Again clamp the instrument by the lower clamp and bring the image of A into coincidence with the vertical hair by means of the lower fine-motion tangent screw. Transit the telescope about the horizontal axis so that the line of sight comes to the position **o**B′ and put a nail in a peg at B′, at the same distance from **o** as the peg B, so that the image of the nail is intersected by the vertical cross hair. Then, as before, angle $B'oC = 2e$ and therefore angle $BoB' = 4e$. Hence, put a nail in a peg at D such that $B'D = \frac{1}{4}B'B$, and, after loosening the diaphragm adjusting screws, move the diaphragm over, being careful to keep the vertical hair vertical, until this hair intersects the image of the nail at D. Tighten the diaphragm screws with the vertical hair in this position, then the line of sight, or line of collimation, should now be at right angles to the horizontal axis. Repeat the test to make certain that the adjustment has been properly made and nothing disturbed while it was in progress.

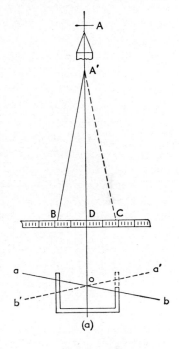

If there is no collimation error, the points B′ and B will, of course, coincide.

The line of sight passing through the centre of the cross hairs and the optical centre of the object glass is often called the "line of collimation". In many ways, however, it is better to confine the term to a line of sight which passes through the cross hairs and the optical centre of the object glass, and at the same time is perpendicular to the horizontal axis; in what follows we shall use the term *line of collimation* in this sense.

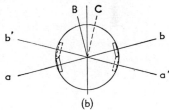

Fig. 4.32.—Horizontal axis adjustment

Horizontal Axis Adjustment.—The object of this adjustment is to make the horizontal axis perpendicular to the vertical axis.

Set up and level the instrument on flat ground near a church tower or other high object, and, with circle left, sight some well-defined point, such as the top of a vane, near or on the top of the object. Let A in fig. 4.32a be the object sighted, and let **aob** be the horizontal axis of the instrument. With both upper and lower plates clamped, depress the objective end of the telescope to sight a point B on a levelling staff placed horizontally some distance away parallel to, and at about the same elevation as, **aob**. Note the reading on the staff.

Change face to circle right and again sight A, clamping both the upper and lower plates with the line of collimation set on A. Depress the objective end of the telescope until the latter sights the levelling staff and take the reading on the staff where the line of collimation intersects it. If the horizontal axis is in adjustment, the reading will be the same as before. If the two readings are different, the mean value should be taken and the line of collimation set to this reading by means of the tangent screw of the upper horizontal plate. Raise or lower the adjustable end of the horizontal axis until the line of collimation intersects the point A when the telescope is raised, using for this purpose the adjusting screws in the standard, underneath the bearing at the end of the axis, which are shown in fig. 4.7 (p. 66). The line of collimation should then intersect both the point A and the middle reading on the staff as the telescope is rotated about its horizontal axis.

The principle of this adjustment will be understood from fig. 4.32, in which the traces, A'B and A'C, of the line of collimation on a vertical plane through the levelling staff are shown in the upper diagram, fig. 4.32a. The lower diagram, fig. 4.32b, represents the two positions of the horizontal axis in plan, and it will be seen that, unless the instrument is in adjustment, these two positions do not coincide but differ by a small horizontal angle.

This adjustment is mainly of importance when horizontal angles have to be observed between points of very different elevations, a case which often arises in mining work or when the legs of the angle are very short. Although the adjustment is one which should be tested at intervals, it should seldom be necessary to alter it, as, once it is made, it is usually not readily disturbed, and reliable makers are careful to send out their instruments with the horizontal axis properly adjusted.

Adjustment of the Telescope Level.—In many theodolites there is no level on the telescope, this being replaced by one on the vernier or microscope frame, but in others there is a level on top of the telescope and none on the vernier frame. When there is a level on the telescope, and vertical angles have to be observed or the instrument used as a level, it is necessary to make this level parallel to the line of collimation so that the latter is truly horizontal when the bubble of the level is at the centre of its run.

The method of testing and making this adjustment is exactly the same as the "two peg" method of adjusting a dumpy level, which is described on p. 125. The adjustment is made by the adjusting

screws at one end of the level. It should, however, be noted that, if
the vernier clip and the vertical circle clamp and tangent screw are on
separate arms on opposite sides of the telescope, this adjustment is not
necessary at this stage as it will be done later in the manner described
on p. 106 as part of the adjustment of the vernier index.

Vertical Circle Index Adjustment.—This adjustment involves making
the vertical circle read zero when the line of collimation is horizontal.
On some instruments the vernier frame is rigidly fixed to the standards,
and no means of making the adjustment for index error is provided.
In this case, all that is possible is to determine the reading on the
vernier when the line of collimation is horizontal, this reading, which
is known as *index error*, being applied, with its proper sign, as a correction to all measurements of vertical angles.

In instruments in which the index is adjustable, the method to
be used will depend on the design of the particular instrument involved. In most modern theodolites, small ones especially, the vernier,
with its index, and the vertical circle tangent screw and clamp are
on separate arms, one on either side of the telescope, while in others,
particularly in some of the older designs, the vertical circle clamp
and tangent screw are on the vernier frame, all on one side of the telescope. In the former case, which is that shown in the theodolite
in fig. 4.7 (p. 66), the clamps and adjusting screws of the vernier
frame and vertical circle are quite independent, and the clip screw
for adjusting the vernier frame is provided with a locking nut, which
must be released before the screw is used and tightened before the
level tube is adjusted. In this type of instrument, the theodolite
fits into its case as a whole, and, once made, the adjustment for index
is permanent in the sense that it does not have to be repeated every
time the instrument is taken out of its box, although, when vertical
angles have to be observed, it is well to test it fairly frequently.

In the theodolite in which the vertical circle clamp, verniers and
fine-motion adjusting screw are on the one arm, the upper part of the
instrument, including vertical circle, horizontal axis and vernier frame,
usually lifts out of the standards and goes into the case separately from
the lower part, including the standards and both of the lower plates. *In
this case the adjustment for index becomes a temporary one, since, when
vertical angles have to be observed, it has to be repeated every time the instrument is taken out of its case and fastened on the tripod.* Hence, the clip of
the vernier frame has no lock nut. When the vertical circle is clamped,
the fine-motion screw moves the circle and telescope relative to the
vernier frame, and, when the vernier frame is clamped, the clip screw

on the latter, when turned, imparts a fine motion of the frame relative to the standards and lower part of the instrument.

The level tube used in vertical-angle observations is generally on the vernier frame, but sometimes it is on the telescope: occasionally, but not often, there is one on both vernier frame and telescope. When the vernier frame is fixed, the level is on the telescope.

We now consider the three cases:

a. Vernier frame not adjustable for index error.

b. The verniers and the vertical circle clamp and tangent screw are on separate arms on opposite sides of the telescope.

c. The verniers and the vertical circle clamp and tangent screw are on the same arm on one side of the telescope.

(*a*) *Vernier Frame not Adjustable.*—In this case we have to determine the index error and use it later as a correction.

1. Make the vertical axis truly vertical by the method described on p. 100, but, instead of the plate levels, use the level on the telescope for the test and the vertical circle tangent screw for correcting half of the movement of the bubble when the instrument is reversed, the other half of the movement being corrected in the usual manner by the footscrews.

2. The bubble on the telescope will then be central, the vertical axis vertical, and, as (p. 104) the level has been made parallel to the line of collimation by the method described on pp. 125-128, the line of collimation will be horizontal. Consequently, the reading on the vernier will be the index error required.

(*b*) *Vernier Clip and Vertical Circle Tangent Screw on Separate Arms.*

1. Level the instrument by the footscrews and set the vertical circle to read zero, using the vertical circle clamp and tangent screw.

2. Note the reading on a levelling staff set up about 100 yards away.

3. Unclamp the vertical circle, transit the telescope, clamp the vertical circle at zero, and turn the instrument through 180° about the vertical axis. Then read the staff again, re-levelling if necessary before taking the reading. If this reading differs from the previous one, set the line of collimation to the mean reading by means of the vertical circle tangent screw and set the vernier to read zero by means of the vernier-frame clip screw. The line of collimation will now be horizontal with the vernier reading zero. Tighten the locking nut of the clip screw and bring the upper bubble, whether on the vernier frame or

on the telescope, to the centre of its run by means of its own adjusting screws, thus making its axis parallel to the line of collimation.

In order to understand this adjustment, suppose that in the first position of the instrument the line of collimation makes an angle α with the axis of the level tube and reads too high. Then, if **R** is the reading of the point on the level staff where a horizontal line from the horizontal axis of the theodolite would meet the staff, the reading on the latter will be $\mathbf{R} + \mathbf{y}$, where $\mathbf{y} = \mathbf{d} \tan \alpha$, $\mathbf{d}$ being the distance from the instrument to the staff. When the telescope is transited and the instrument turned through 180° about its vertical axis to sight the staff again after the vertical circle has been set to zero, the line of collimation will make an angle $-\alpha$ with the axis of the level tube when the bubble is central. Consequently, the reading on the staff will now be $\mathbf{R} - \mathbf{y}$, and the mean of the two staff readings will be **R**. This gives the reading on the staff where a horizontal line of collimation will meet it. Hence, setting the line of collimation to the mean reading on the staff, and setting the vernier to read zero, makes the line of collimation horizontal with zero setting of the vertical circle. This movement will upset the bubble, but bringing it to the centre of its run by means of the adjusting screws on the tube will make the axis of the latter parallel to the line of collimation, which we know is already horizontal. Consequently, after the adjustment has been made, the line of collimation will be horizontal when the bubble is made central and the vertical circle is set to read zero.

(c) *Vernier Clip and Vertical Circle Clamp and Tangent Screw on Vernier Frame.*—Here the adjustment differs slightly according to whether the level tube is on the telescope or on the vernier frame. If it is on the telescope, proceed as follows:

1. Clamp the vertical circle and use the tangent screw to bring the circle to read zero on the vernier.
2. Use the footscrews to bring the telescope bubble to the centre of its run in two positions at right angles to one another.
3. With the telescope bubble in the centre of its run and parallel to one pair of footscrews, turn the whole instrument through 180° about the vertical axis. If the bubble moves from its central position, correct half of the error by the footscrews and half by the clip screw connecting the vernier frame with the standards. The bubble axis will now be perpendicular to the vertical axis, and, as it has already been made parallel to the line of collimation by the method referred to on p. 104, the line of collimation will be horizontal with the vernier reading zero.

If the level tube is on the vernier frame:

1. Level the instrument by means of the plate levels and, by means of its own clamp and tangent screw, set the vertical circle to read zero.

2. Use the clip screw of the vernier frame to bring the bubble on the latter to the centre of its run. Get a labourer to hold a levelling staff about 100 yards away, and note the reading on the cross hair.

3. Unclamp the vertical circle, transit the telescope, set the vertical circle to read zero again, then turn the whole instrument through 180° about the vertical axis to point at the staff, and take the reading on the latter. If the readings on the staff are the same, the adjustment is correct. Otherwise, by means of the vernier-frame clip screw set the line of collimation to the mean reading on the staff and bring the bubble of the level on the vernier frame back to the centre of its run by means of the adjusting screws at one end of the level tube. In this way the line of collimation is first made horizontal with the vernier reading zero and the axis of the bubble is then made parallel to the line of collimation.

Where an instrument has a level tube on both vernier frame and telescope, one or other of the bubbles is used in the above adjustments, and then, when everything is otherwise in adjustment, the bubble of the other tube is brought central by means of its own adjusting screws.

In all of the above described adjustments, the test should be repeated after a first adjustment has been made.

Station Adjustments (i.e. Temporary Adjustments)

Centring.—Centring means bringing the vertical axis of the theodolite immediately over a mark on the ground or under a mark overhead, such as a mark in the roof of a tunnel. In the first case, the string of a plumb bob is attached to a hook under the vertical axis, and the instrument moved until the pointed end of the plumb bob is seen to be exactly over the mark. In the latter case, a plumb bob is suspended under the overhead mark, and the instrument moved until a centring horn on the telescope is directly under the point of the plumb bob.

The first thing to do is to get the instrument centred and levelled as closely as possible by means of the tripod legs: then, after it has been approximately centred, use the centring device to complete the adjustment. As the movement available in the centring device is very limited, the preliminary centring and levelling needs to be reasonably carefully done.

In the centring device consisting of a plate sliding over another plate, the screw clamping ring is loosened and the upper plate slid over the lower one until the instrument is seen to be centred, when the clamp is tightened. The clamping ring may be either above or below the tribrach according to the make of the instrument.

If the centring device is under the trivet stage and consists of two horizontal hinged plates with movements at right angles to one another, the two clamps are loosened and the instrument moved until it is in position, when the two clamping screws are tightened.

Levelling.—Loosen all clamps, and if two level tubes are fitted on the upper plate, the more sensitive of the two should be brought parallel to the vertical plane containing one pair of footscrews, and the bubble brought to the centre of its run by means of these screws. When the screws are being turned, they should be rotated *together* in opposite directions, one clockwise and the other anticlockwise, so that one will tend to raise the low end of the level and the other to lower the high end. When the bubble is central, the second bubble should be brought to the centre of its run by means of the third footscrew or by the second pair of footscrews if the instrument is one with four screws. If there is no second level, the instrument should be turned through 90°, at right angles to the original position, and levelled by the third footscrew or by the second pair of footscrews. Assuming that the level has already been adjusted to be perpendicular to the vertical axis, the horizontal circles should now be level, but this should be tested by seeing if the bubbles remain central while the instrument is turned through a complete revolution about the vertical axis.

When levelling has been completed, it is well to look at the plumb bob to see that levelling has not disturbed the centring.

If vertical angles are to be observed, levelling is best done by the level tube on the vernier frame or on the telescope, as these levels are generally more sensitive than those on the upper horizontal plate.

Focusing.—The method of focusing the telescope has already been described on p. 84. The principal thing in focusing is to eliminate all traces of parallax (p. 85).

23. Adjustment of Instruments with Micrometers or Optical Scales.

The adjustments of the ordinary micrometer microscope have been described in detail on pp. 92 and 93, and need not be repeated here.

Instruments fitted with optical scales and micrometers, however, generally require special adjustments which vary with the make and which are usually described in the instructions issued by the maker. Apart from this, the main adjustments of these instruments are the same as those for an ordinary transit theodolite.

24. Elimination of the Effects of Instrumental Errors by the Use of Special Methods of Observation.

We have mentioned cases where errors arising from faulty adjustment of the instrument may be eliminated by adopting suitable methods of observation, and we can now summarize these cases:

1. Taking the mean of the readings of the two verniers situated at opposite ends of a diameter helps to eliminate the effects of errors of eccentricity due to the axis of rotation and the centre of the graduated arc not coinciding (p. 75).
2. The mean of circle left and circle right observations eliminates the effects of errors of collimation in azimuth and altitude (pp. 68, 69), and of horizontal-axis dislevelment in the case of the measurement of horizontal angles but not for the measurement of vertical angles.
3. Change of swing between observations helps to eliminate errors due to friction and backlash in the moving parts (p. 68).
4. The three-tripod system of observing helps to eliminate the effects of centring errors (p. 96).

There is also one source of error which is hardly appreciable in work with ordinary small theodolites, but which is of importance in geodetic work and which, in the case of horizontal angles, may be considerably reduced, if not eliminated, by using a special method of observing. This error is the error in measurement arising from errors of graduation of the circle. These graduation errors tend to be periodic, and in good geodetic theodolites they are very small, seldom exceeding more than one or two seconds of arc. They are counteracted by observing an angle a number of times, but on different *zeroes*, suitably and evenly spaced on the circle, so that the angle is read several times, each time on a different part of the circle. Thus, with an angle measured 8 times, the initial settings of the circle when the telescope is pointed to the first station would be approximately 0°, 22° 30', 45°, 67° 30', 90°, 112° 30', 135°, and 157° 30'.

There is also another method of observing horizontal angles which, although it does not completely eliminate the effects of small errors of reading either by verniers or micrometers, tends to reduce them considerably. This is the method known as the *method of repetition*, and it consists in measuring some whole multiple of the angle, the value of the latter being a corresponding fractional part of the angle measured on the circle of the instrument.

To measure an angle by repetition, measure it once in the ordinary way but do not disturb the setting when the instrument is clamped pointing to the second station. Loosen the lower clamp, keeping the upper clamp tight, and turn the instrument to sight the first station, bringing the cross hairs into coincidence with the image of the station by means of the lower clamp and tangent screw. The instrument will then be pointing to the first station, but the reading on the circle will be the original reading to the second station. Now loosen the upper clamp and sight the second station, using the upper clamp and tangent screw to bring the cross hairs into coincidence with the image of the station. If the vernier originally read zero at the first pointing to the first station, the reading on the circle (which need not yet be recorded) will now be twice the true value of the angle. Again keep the upper circle clamped and use the lower clamp and tangent screw to bring the line of collimation to sight the first station, after which the previous operation should be repeated. In this way the angle may be repeated on the circle as many times as are desired, and a mean value obtained by dividing the angle read on the circle by the number of sights taken to the second station.

This method of observing angles is particularly useful when very small angles have to be observed. It may not reduce errors of pointing or sighting, but, since the verniers are only read once for each station, and the observed angle is divided by the number of repetitions, the errors of reading (or of graduation) are correspondingly reduced.

25. Accessories for Theodolites.

There are various accessories which are, or can be, used with the theodolite and which deserve mention.

1. *Tripods.*—The tripod of a theodolite consists of three legs of wood or metal joined at the top to a special fitting to take the theodolite; they can be opened out at convenient angles to form a three-legged support. Tripod legs can either be solid lengths of wood or they can be of open-frame construction designed for lightness. Usually the legs of tripods intended to be used with theodolites are in one length, but sometimes they are in two or three lengths which can be slid up to lie together, or alongside each other, when the tripod is not in use. Extensible tripods are convenient for transport, but they are generally not so stable as the ordinary kind and hence are not greatly favoured for anything but the lightest instruments. At the

same time, tripods with sliding legs are useful, if not absolutely necessary, for work in constricted spaces, as in some kinds of mining work.

Certain makers provide a tripod with a *quick-levelling head* which
is exceedingly useful. This consists of a large ball-and-socket arrangement carried on the top of the tripod on to which the theodolite can
be screwed. The arrangement is provided with a clamping ring so
that it can be clamped and held fixed when the theodolite is roughly
levelled, final levelling being done by means of the footscrews in the
ordinary way. The quick-levelling head is best used with small instruments.

The bottom ends of tripod legs are invariably fitted with metal-
pointed *shoes*. These not only protect the end of the leg, but also
enable the end to be driven a short distance into anything but the
hardest ground. For this purpose, the shoe is made with a projecting
lug or lip on which the foot can be placed to help drive the leg home.

2. *Wall Tripods.*—Sometimes it is necessary to set the theodolite
on a staging or flat surface, such as the top of a concrete pillar, where
the ordinary tripod cannot be used. A *wall tripod* is useful in such
cases. This consists of a plate on top of which there is a threaded collar,
on to which the theodolite can be screwed; the plate is supported by
three short pointed legs about two or three inches long.

3. *Case and Outer Cover.*—The ordinary theodolite is packed in a
strong rectangular wooden case, with shaped partitions and stops in
which the instrument fits snugly and is held firmly without rattling.
When working with a new instrument, its exact position in the case
should be very carefully noted, so that it can be replaced easily and
without strain. All clamps should be released before the theodolite
is put in the case, and care should be taken to see that it is placed
very gently in position without dropping or straining it.

In order to protect the instrument against moisture, outer cases
of leather or waterproof canvas are obtainable and should always be
used. These cases are often provided with special shoulder straps to
enable the instrument to be carried on the back or loosely over one
shoulder.

In late years the wooden case is often replaced by one of metal,
and the instrument is placed in the case and either screwed to the
bottom or held down firmly in an upright position. These metal cases,
which are generally waterproof, are usually lighter and more convenient than wooden cases, and they hold the instrument more firmly
and with less chance of strain.

4. *Compasses.*—Theodolites are often fitted with compasses; if they

are not, they generally have attachments which allow for a compass of some sort to be fitted as an extra later on if required.

What used to be the most common form was an ordinary compass, with graduated ring in a circular box, fitted on the upper plate between the standards. This form is still provided on some smaller instruments.

Trough compasses, either fitted to one of the plates or attached to one of the standards, are also met with occasionally, but most modern theodolites, if provided with a compass at all, are fitted with one of the tubular variety. This is generally fitted to one of the standards.

5. *Internal Illumination.*—If a theodolite is to be used at night, or in dark places, it is necessary to provide some sort of illumination of the cross hairs and verniers. Before electric dry batteries were brought to their present state of perfection and reliability, the most usual form of illumination of the cross hairs was a small oil lamp fitted in a bracket carried by one standard. One end of the trunnion axis had a horizontal hole running through its length, and the light from the lamp was made to shine through this hole to a small mirror in the telescope barrel, from whence it was reflected on to the cross hairs.

Another form of illumination which is still used occasionally is a small reflecting prism fitted in the sun cap at the objective end of the telescope. An electric torch or oil lamp is held close to the prism, and the light is reflected through the objective in the direction of the cross hairs.

With the above-described methods of illuminating the cross hairs no special provision was made for illuminating the graduated arcs, and these were read by means of an oil lamp or electric torch held by hand at the side of the instrument. The modern theodolite with glass circles, however, requires good illumination of the circle and micrometers, and, for night work, this is generally provided as an extra fitting by small electric lamps built into the instrument and worked by dry batteries or accumulator, the cross hairs being illuminated in the same way. In the day time, natural light is reflected by mirrors or prisms into the glass circle and thence into the micrometers.

6. *Striding Level.*—A striding level often forms part of the equipment of moderately sized and large theodolites. This is a specially sensitive level, used mainly in astronomical observations for levelling the horizontal axis, or for measuring any error in the level of this axis. It is mounted on two long legs, one on either end of the level tube, which fit over the ends of the horizontal or trunnion axis. A striding

level is seldom used in ordinary work, and small theodolites are normally not provided with one.

7. *Optical Plummet.*—An optical plummet is an optical substitute for the plumb bob used in centring the instrument, and can only be obtained with certain models. In some instruments, as in the Watts " Microptic " Theodolite No. B50, it is built into the theodolite and forms part of it. In others it consists of a small telescope which fits into the levelling head. The telescope is provided with cross hairs which give a line of sight parallel to, and coincident with, the vertical axis of the instrument.

An optical plummet can be tested by setting up another theodolite a short distance away from the one being tested in two positions at right angles to one another. After the two theodolites have been carefully levelled, the line of collimation of the extra theodolite, when directed at the centre of the one under test, should, when depressed slightly, intersect a mark on the ground previously established by means of the optical plummet. After a test has been completed in one position, another should be made from the second position at right angles to the first.

8. *Diagonal Eyepiece.*—The diagonal eyepiece is used for astronomical observations, and also for observations to very high objects which cannot be observed in the ordinary way because of the lower part of the theodolite getting into the way of the ordinary eyepiece. The diagonal eyepiece is an eyepiece fitted with a reflecting prism to reflect the rays of light at right angles to the ordinary line of sight. Hence, when the telescope is nearly vertical, the eye end of the diagonal eyepiece is almost horizontal. When the diagonal eyepiece is to be used, the ordinary eyepiece is removed and replaced by the diagonal eyepiece.

9. *Dark Shades.*—These are very dark-coloured glasses which can be slipped over the eyepiece to protect the eye during observations to the sun.

THE MINING DIAL

26. Characteristics.

The mining dial is an instrument, much used in mining surveying, which combines in its construction a fairly large surveyor's compass with what is in effect a simple form of theodolite.

Fig. 4.33 shows an improved type of mining dial manufactured by Messrs. Hilger and Watts, Ltd. It has the ordinary levelling head and upper and lower plates of a theodolite, and also telescope and vertical circle, but no centring device. The horizontal circle, which reads to 1' by means of two oppositely situated verniers, is $4\frac{3}{4}$ in. diameter, and the vertical circle, which reads to 2' by means of two verniers, is 3 in. in diameter. The compass needle is 4 in. long, and the zero of

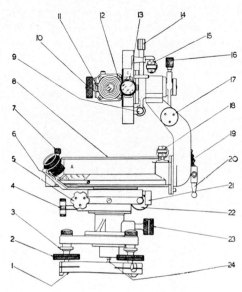

Fig. 4.33.—Mining dial by Messrs. Hilger and Watts, Ltd.

1. Footscrew ball adjusting screw.
2. Footscrew.
3. Footscrew adjusting nut.
4. Lower clamp.
5. Lower tangent screw.
6. Horizontal circle vernier.
7. Compass needle raising knob.
8. Glass cover to compass bowl.
9. Altitude bubble tangent screw.
10. Telescope focusing milled head.
11. Graticule illumination window.
12. Graticule adjusting screws.
13. Vertical circle vernier.
14. Altitude bubble clamp.
15. Altitude bubble collets.
16. Telescope clamp.
17. Telescope tangent screw.
18. Plate bubble collets.
19. Locating pin.
20. Telescope bracket clamp lever.
21. Upper clamp.
22. Upper tangent screw.
23. Levelling base clamp.
24. Plumbing hook.

(By courtesy of the makers)

the graduated arc in the compass box is set to be parallel to the line of collimation of the telescope. The tripod is fitted with a ball-and-socket quick-levelling device, by which, and by using a round bubble in the compass box, the instrument may be roughly levelled, the final levelling being done in the usual manner by the footscrews and the level on the horizontal upper plate.

This particular instrument possesses two features not to be found on the ordinary simple mining dial. One is that the arm carrying the vertical circle and telescope can be released and turned about a horizontal axis through an angle of about 45° and then clamped. This allows the telescope to be used for vertical, or almost vertical, sights. The other is that the dial can be interchanged with special lamp cups on the levelling head. These cups are so designed that the flame of a standard Davy lamp is always in the centre of the cup. By this means, the dial can be used with lamps for the three-tripod system of observing (p. 96).*

The *adjustments of the mining dial* are in general similar to those of the theodolite and surveyor's compass.

* *Note.* Messrs. Hilger & Watts have since discontinued the manufacture of the mining dial described above and have replaced it by two types of light theodolite with the telescope supported as in fig. 4.33 on one standard only and with circles reading direct to 5′ and by estimation to 1′. A tubular compass supported on the top of the standard is obtainable as an extra and thus replaces the large surveyor's compass used in the mining dial.

CHAPTER V

INSTRUMENTS FOR MEASURING SLOPES, DIFFERENCES OF ELEVATION AND HEIGHTS

1. Main Classes of Instruments.

In this chapter we shall be concerned with instruments for measuring slopes and vertical angles, differences of elevation, and absolute heights above sea-level. For this purpose, we shall divide these instruments into the following classes:

1. Clinometers and hand levels.
2. Surveyor's level.
3. Barometers, aneroids and hypsometers.

Clinometers and *hand levels* are very much lighter, more compact and simpler in use than the surveyor's level, and they are only used when moderate accuracy is needed, such as in measuring slopes for chainage corrections or spot heights for contouring. The first measures slopes or vertical angles, and the second measures differences of elevation.

The *surveyor's level*, generally simply called a *level*, is used for all work in which elevations have to be accurately determined, or when special levels and slopes have to be accurately and carefully set out, and it is one of the most extensively used of all survey instruments. Thus, a level would be used for measuring the fundamental heights in connexion with a large contoured survey, for setting out levels for excavations for dams, bridge abutments, &c. A theodolite can be used to measure vertical angles, and, although not so convenient for the purpose, it can, if necessary, be employed for all work that can be done with a level; but a level cannot be used for all the work that is possible with a theodolite.

The *surveyor's level* measures differences of height, but *barometers*, *aneroids* and *hypsometers* may be used for the direct measurement of elevations above some standard elevation—in most cases mean sea-level. The readings on the barometer, however, are only approximate, even after they have had all necessary corrections applied to them, when compared with results obtained by ordinary levelling. For this reason, the chief function of the barometer and aneroid is for determinations of heights, or differences of heights, for exploratory and small-scale mapping purposes. The hypsometer, or boiling-point thermometer, is also a comparatively rough instrument mostly used for exploratory work.

CLINOMETERS AND HAND LEVELS

2. Clinometers.

Clinometers are instruments, usually held in the hand, for measuring slopes or angles of slope. They are of various types, but consist essentially of a graduated arc against which the slope is read, either with reference to a freely suspended plumb bob or else to an index mark on an arm carrying a spirit level placed at right angles to the line joining the index to the centre of rotation of the arm.

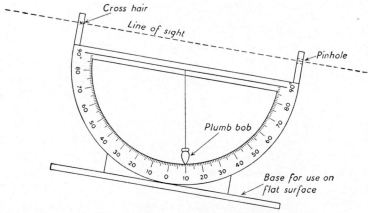

Fig. 5.1.—Diagram illustrating clinometer

Fig. 5.1 shows diagrammatically a simple form of clinometer with pinhole sight and cross hair on top of a graduated arc. In this particular case the flat base can be used to lay the instrument on a smooth surface whose slope is required.

A clinometer is sometimes embodied in the lid of a prismatic compass case. In one form of this type, a graduated arc is pivoted to the compass case and is weighted so that the zero mark always hangs vertically under the pivot when the arc is allowed to swing freely in a vertical plane. A line etched on a glass window in the compass case serves as an index to which the readings on the graduated arc are referred. This form of clinometer is much favoured by prospectors and geologists for measuring the angle of dip and direction of strike of strata, the compass being used for ordinary survey work when magnetic bearings are needed. The index line is at right angles to the ordinary line of sight so that the latter can be used when measuring the angles of slope of long lines, but for such work as measuring the dip of strata, the instrument is placed direct on the surface whose dip is to be measured, special feet with flat sides parallel to the ordinary line of sight being fitted underneath the index window.

3. Hand Level.

The hand level, as its name implies, is a small level held in the hand, and consists of a small rectangular metal box, open at one end,

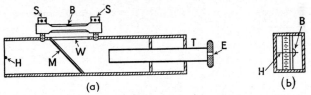

Fig. 5.2.—Diagrammatic section of hand level

and fitted at the other with a round metal tube which can be drawn in and out. The outer end of this tube is closed with a cap in which a peep-hole, E in fig. 5.2a, has been bored. On top of the box is a rectangular opening W, and immediately above this opening is a level tube B in which the lower and upper surfaces of the level are clear. Inside the metal box, and to one side of it immediately under the window, is a nickel-silver mirror M set at an angle of 45° to the main length of the box. This mirror only occupies one half of one side of the box, the other half of the latter being left clear. At the open end of the box is a horizontal wire H, which, with the peep-hole E, gives a definite line of sight.

The level is used by holding the peep-hole E close to the eye and sighting towards a levelling staff or target. This staff or target is seen direct in one half of the field of view. When the level is held horizontal, an image of the bubble, reflected in the mirror M, is seen in the other half of the field of view (fig. 5.2b), and the observation consists in noting the reading of the staff against the cross hair when the centre of the bubble appears to lie against the cross hair as shown in fig. 5.2b.

The adjustment of the hand level consists in making the level bubble parallel to the line of sight. For this purpose, the two-peg method of adjusting the surveyor's level (pp. 125–128) is adopted, the actual adjustment of the level tube being made by means of the adjusting screws S (fig. 5.2a) at the ends of the tube.

4. The Abney Level.

The Abney level (fig. 5.3, Plate IX) can be used either as an ordinary hand level or as a clinometer. It consists of a hand level in which the bubble is fixed on top of a rotatable arm, the lower end of the arm terminating in an index and vernier. This arm works against a graduated arc carried on one side of the rectangular box containing the mirror. A small magnifying glass is usually attached to enable the vernier to be read easily and quickly. In some levels the arm is rotated by means of a round milled head carried at one end of the axis, but in others it is rotated by means of a rack and pinion operated by a fine-motion tangent screw.

In observing slopes, a pole or staff, with a target set above the bottom of the staff at the same height as the eye, is held at the point where the sight is to be taken. The level is then directed to the target so that the cross hair intersects the latter, and at the same time the milled head or tangent screw is turned to bring the bubble, as seen by its reflected image, to the centre of its run opposite the hair. The reading on the vernier then gives the slope required.

If the instrument is to be used as a hand level, the vernier is set to read zero on the graduated arc and the level is then used as an ordinary hand level.

Abney levels are sometimes graduated to give actual slopes, but more often they are graduated in degrees, with a vernier reading to 10′ or 5′. Occasionally an Abney level is graduated both in slopes and degrees.

The adjustment of the Abney level can be carried out by setting it to read zero and then adjusting it as an ordinary hand level. Alternatively, it can be used to read the same slope in reverse directions. If both readings are the same, no adjustment is necessary. If the readings differ, set the instrument to read the mean of the observed slopes, then sight the slope in one direction and use the adjusting screws of the level tube to bring the bubble to the centre of its run as the sight along the slope is taken.

PLATE IX

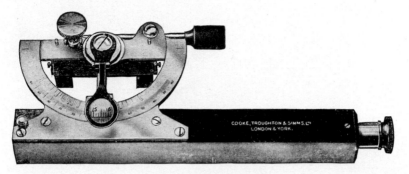

Fig. 5.3.—ABNEY LEVEL
(By courtesy of Messrs. Cooke, Troughton and Simms, Ltd.)

Fig. 5.6.—DUMPY LEVEL
(By courtesy of Messrs. W. F. Stanley & Co., Ltd.)

THE SURVEYOR'S LEVEL

5. General Description and Types of Level.

The surveyor's level is a much larger instrument than the hand level and also, of course, as it is always used mounted on a tripod, it is much more accurate. It consists essentially of a level tube mounted on a telescope provided with suitable cross hairs and supported on a levelling head, somewhat similar to the levelling head of a theodolite but without a centring device. The bubble of the level tube defines a horizontal plane, and, when the instrument is in adjustment, the line of sight which passes through the centre of the cross hairs and the optical centre of the telescope is parallel to this plane. Hence, when the bubble is central the line of sight should be horizontal. Levelling consists in measuring the vertical heights of a continuation of the line of sight above some definite points on which (while observations are in progress) a graduated level staff is placed and held vertical.

There are two main types of surveyor's level—the *dumpy level* and the *wye level*—and we shall also describe two other types—the *tilting* and *reversible* or *self-adjusting levels*—which are modifications of the two main types and which are now very extensively used. All surveyor's levels, however, depend ultimately for their working on the ordinary level tube containing a bubble of air, or vapour, and a suitable liquid. Accordingly, we shall begin with a description of the construction and principles involved in this piece of apparatus.

6. The Level Tube.

The level tube consists of a glass tube, partially filled with liquid, the inner surface of which is carefully ground so that a longitudinal section of it by a vertical plane through the axis of the tube is part of a circular arc. If the tube is always to remain in the same position on the instrument, only the top half of the section need be circular, but if the tube is sometimes also to be used upside down, as in the case of the reversible level, both top and bottom halves of the section must be arcs of circles of opposite curvature, so that the tube resembles an elongated barrel. Before it is sealed, the tube is partially filled with a liquid of low viscosity, such as alcohol, chloroform or sulphuric ether, leaving a small space which forms a bubble of mixed air and vapour. Under the action of gravity, the bubble will always rise to the highest part of the tube, and thus comes to rest, as in fig. 5.4, so that a tangent plane to the inner surface of the tube at the highest point of the bubble defines a horizontal plane.

The outer part of the tube is usually graduated with a short series of equally spaced graduations, or else two marks are placed where the ends of the bubble come when it is at normal temperature, and the bubble is considered to be central when it is symmetrically spaced with respect to these lines. The sensitiveness of a bubble is usually

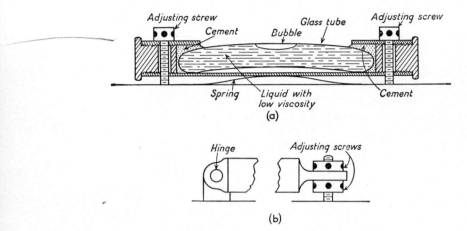

Fig. 5.4.—Diagrammatic section of a spirit level

defined as the angular value of one division, the relation between them being

$$\theta = \frac{d}{R} \times 206265,$$

where θ is the value in seconds of arc of one division, d the distance between two consecutive graduations, and R the radius of the circular arc of a longitudinal section of the curved surface expressed in the same linear units as d. Here, of course, the factor 206,265 is the number of seconds of arc in one radian.

The bubble tube is contained in a cylindrical metal tube, open at the top so that the bubble and the graduations on the glass can be seen easily. When the level tube is on an instrument where it can be reversed, so that the lower side of it sometimes becomes uppermost, the lower side of the metal tube is also open to permit of the bubble and graduations being seen in the reverse position. The bubble tube is fixed in the metal tube by plaster of Paris, and, if a bubble tube gets broken, it should be taken out, all the solid plaster of Paris scraped away, and a new tube put in and fastened by means of fresh plaster mixed with a little water.

In certain levels the graduations are done away with, and an optical arrangement enables the two ends of the bubble to be viewed side by side as in fig. 5.5a. The bubble is central when the two ends appear to coincide, as in fig. 5.5b.

The level tube is usually provided with a means of adjustment by means of which one end can be lowered or raised. In fig. 5.4a, capstan adjusting screws are fitted at both ends, and the tube can be adjusted by loosening one screw and tightening the other.

In large instruments, however, such as the theodolite and surveyor's level, one end of the tube is hinged and the other end works up and down a vertical screwed post provided with capstan tightening nuts, one above and one below the end of the tube, as in fig. 5.4b.

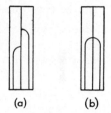

(a)   (b)

Fig. 5.5.—Diagram of bubble ends

In the ordinary level tube, the length of the bubble varies with temperature, becoming shorter as the temperature rises. This at times may become inconvenient, but Messrs. Hilger and Watts, Ltd., have patented a " Constant " level tube in which the length of the bubble remains constant under varying conditions of temperature.

The value of one division of a bubble tube can be found by sighting a level staff and noticing the difference in readings on it as the tube is tilted so that one end of the bubble moves through n divisions on the scale. Let r be the difference in reading on the staff, l the distance of the staff from the level expressed in the same units as r, θ the value of one division on the scale expressed in seconds of arc, n the number of divisions through which the bubble has moved. Then,

$$\theta = 206265 \times \frac{r}{nl}.$$

Also, if d is the length of one division of the scale, expressed in the same linear units as r and l, and R is the radius of the tube expressed in the same linear units,

$$R = \frac{d \cdot n \cdot l}{r}.$$

This formula follows from the fact, which can easily be seen by drawing a simple diagram, that the angle subtended by the difference of readings on the staff is $2 \tan^{-1} \dfrac{r}{2l}$, which, as the angle is small, may be written as $\dfrac{r}{l}$ radians. This angle represents the difference between the two positions of the line of collimation as the level is tilted. But this angle is also the angle through which the bubble has moved and which measures $\dfrac{nd}{R}$ radians. Hence,

$$\frac{nd}{R} = \frac{r}{l}, \text{ or } R = \frac{d \cdot n \cdot l}{r}.$$

7. The Dumpy Level.

The dumpy level (fig. 5.6, Plate IX, p. 120) consists of a large-aperture telescope carried on a vertical spindle similar to the spindle of the upper circle of a theodolite. This spindle works in a vertical bearing forming part of a levelling head, also very similar to the levelling head of a theodolite but somewhat lighter and without a centring device. In a three-screw instrument the levelling head consists of the usual tribrach with three levelling footscrews and a trivet stage which can be screwed on to a tripod. A sensitive level tube is fitted on top or on the side of the telescope, or else on a horizontal bar or stage carrying collars on top to hold the telescope and with the vertical spindle underneath. Thus, in the dumpy level there is only one axis of rotation—the vertical axis.

The telescope is somewhat longer than the telescope of the theodolite but is otherwise very similar. The size of a level is defined by the length of the telescope in the case of an internal focusing instrument, or by the focal length of the objective in the case of an ordinary telescope, and this length varies from about $6\frac{1}{2}$ in. to 16 in. A large-diameter objective, not less than $1\frac{3}{8}$ in., is desirable. The cross hairs normally consist of a single vertical and horizontal hair as shown in fig. 4.16a (p. 83), but sometimes stadia hairs are also fitted as in fig. 4.16d. The intersection of the cross hairs and the optical centre of the objective give a line of sight which, when the instrument is in adjustment, is perpendicular to the vertical axis and parallel to the tangent at the highest point of the curve of the bubble when the latter is central. When these conditions are satisfied, the line of sight is called the *line of collimation*.

In some cases the instrument is fitted with a clamp and fine-motion tangent screw enabling the telescope and its spindle to be clamped to the tribrach and a fine, horizontal motion given to them relative to the latter; but in many cases no clamp or tangent screw is provided.

8. Adjustments of the Dumpy Level.

The adjustments of the dumpy level are:

Permanent Adjustments

1. Adjustment of the level tube, i.e. to make the level bubble perpendicular to the vertical axis so that the bubble is at the centre of its run when the vertical axis is truly vertical.
2. Adjustment of the line of sight, or to make the line of sight of the telescope parallel to the level bubble.

Station Adjustments

1. Levelling up, or making the vertical axis vertical.
2. Focusing the telescope.

Permanent Adjustments

Adjustment of Level Tube.—This adjustment is exactly the same as the adjustment of the levels on the upper plate of a theodolite. After levelling up roughly, set the level tube parallel to a line joining one pair of footscrews and bring the bubble to the centre of its run by means of the footscrews. Reverse the instrument by turning it through 180° about the vertical axis and see if the bubble is still central. If not, correct half the error by the footscrews and the remaining half by the adjusting screws on the level tube. Repeat the test over the same pair of footscrews and then in a position at right angles to this.

Adjustment of Line of Sight. Two-peg Test.—Lay out two pegs A and B about 150 ft. apart (fig. 5.7a, p. 126) and set the instrument with the telescope pointing to B and the eyepiece E vertically over the peg A. After levelling up the instrument carefully, put a levelling staff on A and observe the height a of the centre of the eyepiece above the peg A. Next get an assistant to hold a staff on B, and, after careful focusing (p. 84), note the reading b on the staff. If the instrument is in adjustment, the difference of height between A and B will be BC − AE, where EC is a horizontal line through E meeting the staff on B at C, and BC and AE are the readings of the staff at B and A. If the instrument is not in adjustment, the line of sight will not pass through C. Suppose it meets the staff on B at a point D a distance e below C. Then, $BC = b + e$. Hence, the true difference of height between A and $B = (b + e - a)$. Now move the instrument and set it up near B so that the eyepiece is over B (fig. 5.7b). With the levelling staff set on B, measure the height b' of the centre of the eyepiece above B, and, after carefully levelling and focusing, take the reading a' on the staff when the latter is moved and set up on A. Then it is easy to see that the true difference of height between B and A = $b' - (a' + e)$. Hence, we have:

$$b + e - a = b' - a' - e,$$

or,

$$e = \tfrac{1}{2}[(b' - a') - (b - a)].$$

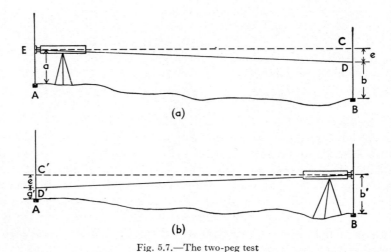

(a)

(b)

Fig. 5.7.—The two-peg test

The term in square brackets is the difference between the two apparent differences of elevation, and it will be seen that, if the line of sight were truly horizontal, the reading on the staff at A would be

$$a' + \tfrac{1}{2}[(b' - a') - (b - a)].$$

Hence, keeping the instrument at B, the adjustment consists in bringing the line of sight to this reading by means of the adjusting screws of the telescope diaphragm.

The method just described is, perhaps, the most commonly used method of adjusting the line of sight, but the following one is to be preferred, especially for use with internal focusing telescopes, or if there is any possibility of the objective end of the telescope drooping.

Set out two pegs A and B as before, but set the instrument up at a point exactly half-way between them (fig. 5.8a). After careful level-ling and focusing, sight on a staff held on A and take the reading a. If the line of sight is truly horizontal, the reading will give the true distance of the peg below a horizontal line through the instrument. Let the line of sight not be truly horizontal when the bubble is at the centre of its run and let it intersect the staff at a distance e below the horizontal line through the instrument. Then the true distance of the peg below the latter line will be $(a + e)$. Now set the staff

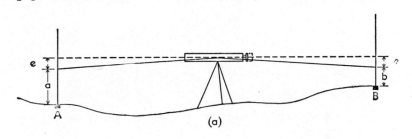

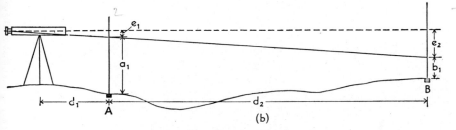

Fig. 5.8.—Alternative method of collimation adjustment

over peg B and note the reading b. As the level is equidistant from the pegs A and B, the error in reading on the staff will also be e, and the distance of B below the horizontal line through the instrument will be $(b + e)$. Hence, the difference in height between A and B will be $(a + e) - (b + e) = (a - b)$, or the difference between the actual readings at A and B. Consequently, we have eliminated the effects of the error of collimation and found the true difference in level between the two pegs. Call this difference h.

Remove the instrument and set it a short distance d_1 from A, the distance between the two pegs being d_2. Read a staff held on A and then on B, and let the readings be a_1 and b_1 respectively. Owing to the inclination of the line of sight to the horizontal, these readings will be in error by amounts e_1 and e_2, and from fig. 5.8b it can easily be seen that the true difference in height between A and B will be $(a_1 + e_1) - (b_1 + e_2) = (a_1 - b_1) + (e_1 - e_2)$, $(a_1 - b_1)$ being the apparent difference in height. But, from similar triangles,

$$e_1 - e_2 = -\frac{e_1 d_2}{d_1}.$$

Hence, the difference in height between A and B

$$= (a_1 - b_1) - \frac{e_1 d_2}{d_1}.$$

$$\therefore\ h = (a_1 - b_1) - \frac{e_1 d_2}{d_1},$$

whence

$$e_1 = \frac{d_1[(a_1 - b_1) - h]}{d_2}.$$

Consequently, the reading on the staff at A when the line of sight is horizontal will be

$$a_1 + e_1 = a_1 + \frac{d_1}{d_2}[(a_1 - b_1) - h].$$

Hence, if the diaphragm is loosened and moved until the cross hair gives this reading on the staff at A when the bubble is at the centre of its run, the hair will then be horizontal and parallel to the bubble.

If the sight is adjusted to the reading on the staff at B, this reading should be

$$b_1 + e_2 = b_1 + \frac{(d_1 + d_2)}{d_2}[(a_1 - b_1) - h].$$

The signs in these formulæ depend on the sign of the term in square brackets. If the apparent difference in level between the two points, as given by the second observation, is greater than the true difference, as given by the first observation, the correction is additive and the line of collimation must be moved *up*. If the apparent difference of level is less than the true difference, the correction is subtractive and the line of collimation must be moved *down*. The adjustment, of course, must be done by means of the diaphragm adjusting screws.

Temporary Adjustments of the Dumpy Level

Levelling up.—Levelling up a dumpy level is effected in exactly the same manner as levelling up an adjusted theodolite, as described on p. 109, except that the main level tube on the level is used instead of the level tube on the horizontal plate of the theodolite. When the instrument is properly levelled, the bubble in the level tube should remain central in all positions of the telescope as the latter is turned through a complete revolution about the vertical axis.

Focusing the Telescope.—Focusing the telescope is done as described on p. 84, the cross hairs being focused first by the eyepiece, and the image of the staff then brought to the plane of the cross hairs by means of the main focusing screw so that the image is seen to be free from any signs of parallax.

9. The Wye Level.

The fundamental difference between the dumpy and wye level is that in the former the telescope is fixed with relation to the vertical spindle, whereas in the wye level the telescope and vertical spindle are separate, the former being carried in two vertical " wye " supports fixed to a horizontal bar or stage, which in turn is attached to the vertical axis. The telescope can be removed from the wyes or it can be rotated in them about its longitudinal axis and then held fixed in any position by clips forming parts of the wyes. The level tube may be attached either to the telescope, in which case it must be of the reversible type, or else mounted on top of the stage carrying the wyes, the former being the more usual arrangement. One or both of the wyes is adjustable up or down by means of stage adjusting screws at the point where the wyes are attached to the stage.

One form of wye level is shown in fig. 5.9 (p. 130). Each wye consists of a semicircular support shaped to form the lower half of a bearing for a cylindrical collar sweated on to the telescope barrel. On top is a semicircular clip, of the same radius as the wye, which is hinged at one end to one side of the wye and fits at the other into a slot in the other side of the wye. This clip fits tightly over the top of the collar on the telescope and, when closed, can be secured in the slot in the wye by means of a pin. Hence, when the telescope is placed in the wyes and the two clips closed, it is held fixed relative to the stage and to the vertical axis. If the clips are opened, the telescope can

either be rotated about its longitudinal axis or it can be taken out
of the wyes and replaced end for end.

The bottom of the instrument consists of the usual tribrach and
trivet stage complete with levelling footscrews.

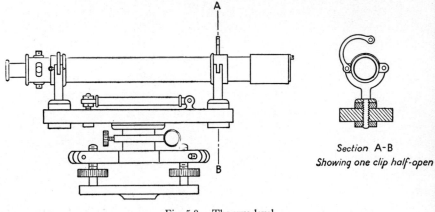

Fig. 5.9.—The wye level

10. Adjustments of the Wye Level.

(a) *Level Tube Attached to Telescope.*

When the level tube is attached to the telescope, the permanent
adjustments of the wye level are:

1. *Adjustment of the Line of Sight.*—To make the line of sight coincide
   with the axis of the collars, so that a rotation of the telescope about
   its longitudinal axis is also a rotation about the line of sight.
2. *Adjustment of the Level Tube.*—To make the axis of the level tube
   parallel to the line of sight.
3. *Adjustment for Perpendicularity of Vertical Axis and Level Tube.*—
   To make the axis of the level tube perpendicular to the vertical
   axis so that the level bubble is central when the vertical axis is truly
   vertical.

Adjustment of Line of Sight.—Set up the instrument and sight, and
carefully focus, some well-defined point not too far away. Loosen the
clips and rotate the telescope through 180° about its longitudinal
axis. Fasten the clips and see if the point still appears to be at the
intersection of the cross hairs. If not, adjust half the error by the diaphragm screws and then repeat the test on a different point.

The reason for this method of adjustment can easily be understood by imagining that the line of sight makes a small angle α with the axis of the collars. When the telescope is rotated through 180° about its longitudinal axis, in this case also the axis of the collars, the line of sight will make angle $-\alpha$ with the latter axis, so that the total apparent movement is 2α. Hence, by adjusting half the difference between the two positions of the line of sight, we bring the latter into coincidence with the axis of the collars.

Adjustment of the Level Tube.—This adjustment is made up of two parts, (1) to bring the axis of the level tube into the plane containing the line of sight, and (2) to bring the axis of the level tube parallel to the line of sight.

For the first part of the adjustment, level the instrument over one pair of footscrews and then turn the telescope slightly in the wyes about its longitudinal axis. The bubble should remain at the centre of its run. If it does not do this, make it central by means of a small *horizontal* adjusting screw which is fitted at one end of the tube for the purposes of this adjustment.

The second part of the adjustment of the level tube can be carried out in either of two ways. One is to set the tube parallel to one pair of footscrews and level it in this position by means of these screws. Then take the telescope out of the wyes, turn it end for end, and replace it in the wyes in this position. The bubble should still be at the centre of its run. If not, correct half the error by the adjusting screws at the end of the tube. Alternatively, the adjustment can be tested by means of the two-peg test used in the case of the dumpy level (pp. 125–128), and, if necessary, corrected by the level tube adjusting screws.

Adjustment for Perpendicularity of Vertical Axis and Axis of Levelling Tube.—The test for this adjustment is carried out in a manner similar to that used in the adjustment of the plate levels of the theodolite, or of the level tube on a dumpy level (pp. 99 and 125), but here any error after reversal of the upper part of the instrument through 180° about the vertical axis is adjusted half by the footscrews, and half by raising or lowering one wye relative to the other by means of the adjusting screws at the base of the wye where it is joined to the stage.

(b) Level Tube on Stage under Telescope.

When the level tube is on the stage of the instrument as in fig. 5.9 (p. 130), the adjustments are slightly different. In this case the first adjustment remains as before but the other two are:

2. *Adjustment for Perpendicularity of Vertical Axis and Level Tube.*— To make the axis of the level tube perpendicular to the vertical axis.
3. *Adjustmeut for Parallelism of the Line of Sight and the Axis of the Level Tube.*—To make the line of sight parallel to the axis of the level tube.

Adjustment No. 2 is done in the usual way by reversing the instrument through 180° about the vertical axis after it has been accurately levelled over one pair of footscrews. Half of the error on reversal is corrected by the footscrews and half by the adjusting screws at the end of the level tube.

The last adjustment is made most easily by sighting a staff with the instrument carefully levelled and taking a reading. Then reverse the telescope end for end in the wyes and again sight and read the staff. If the readings are different, bring the line of sight to the mean reading on the staff by means of the adjusting screws under one wye.

11. The Tilting Level.

The tilting level has the advantage that, for running straight lines of levels, it need only be levelled approximately by means of the footscrews, final levelling for each individual sight being done by means of a screw situated under the eyepiece end of the telescope. Turning this screw causes the telescope and attached level to be tilted together relative to the base of the instrument about a horizontal hinge below the objective end of the telescope. Hence, when a sight is being taken, the level must be approximately levelled by the footscrews with the aid of a circular level attached to the stage, and the bubble brought to the exact centre of its run by means of the tilting screw. This tilting screw sometimes carries a graduated drum working against an index, the graduations on the drum indicating different gradients.

Fig. 5.10, Plate X, shows the " High-power " Quick-setting Tilting Level manufactured by Messrs. Hall Bros. (Optical), Ltd. This instrument is not provided with the ordinary tribrach and levelling footscrews, approximate levelling being done instead by means of a ball-and-socket rough-levelling arrangement and a circular level. The main level tube (not visible in the illustration) is mounted at the side of the telescope and is fitted with a hinged mirror which reflects an

PLATE X

Fig. 5.10.—THE TILTING LEVEL
(By courtesy of Messrs. Hall Bros. (Optical), Ltd.)

Fig. 5.14.—ANEROID
BAROMETER
(By courtesy of Messrs. Hilger
and Watts, Ltd.)

(G 446)

image of the bubble to the observer as he stands at the eyepiece end of the telescope, thus avoiding his having to move to one side to observe the bubble. The instrument is provided with a graduated horizontal circle, and a vernier reading to 5′ of arc.

The tilting level is an exceedingly convenient instrument when only a very few sights have to be taken at each set-up, but it has the disadvantage that, when a number of sights have to be taken at the one set-up, as in taking spot heights for contouring, it has to be levelled for every sight. When an ordinary dumpy or wye level is in adjustment, a number of sights can be taken in different directions with a single levelling. A great advantage of the tilting level, however, is that, with the reflecting mirror, the bubble can be seen and its setting accurately and quickly adjusted at the same time as the staff is being read. For this reason, the tilting principle is now generally embodied in levels intended for the most precise work.

12. Adjustment of the Tilting Level.

The only permanent adjustment necessary is to ensure that the line of sight or collimation is horizontal when the bubble is at the centre of its run. The method of carrying out this adjustment is the two-peg method used in the adjustment of the dumpy level (pp. 125–128). Having determined the graduation on the staff which the instrument should read when the line of sight is horizontal, direct the cross hair to this reading by means of the tilting screw. Unless the instrument is in adjustment, the bubble will no longer be in the centre of its run. Bring it back to the centre by means of the level adjusting screws at the end of the level tube.

13. The Reversible Level.

The reversible level, often called the *self-adjusting level*, combines some of the features of the wye level with those of the dumpy level. The level tube in this instrument is generally at the side of the telescope, the latter being rotatable about its longitudinal axis as in the wye level, so that it can be turned over about this axis and used with the level tube on either side of the observer as he looks through the eyepiece. In this case, of course, the level tube is open on its top and bottom sides in order that the bubble and the graduations on the bubble tube can be seen in whichever position the telescope happens to be. The advantages of the reversible level are that collimation adjustment is very easy and the mean of two readings, the one with the level tube on one side of the instrument and the other with it on the other side, gives a result from which errors arising from faulty collimation adjustment are eliminated.   The instrument is almost invariably constructed on the tilting principle, and is therefore used exactly like the ordinary tilting level, except that in accurate work readings are taken with the level tube in the two positions, one with level right and the other with level left.

Nearly all modern instruments intended for precise work or geodetic levelling are combined tilting and reversible levels.

14. Adjustment of the Reversible Level.

The adjustment of the reversible level consists in making the line of sight or collimation parallel to the axis of the level tube and is made in two stages. The first stage is to make the line of sight horizontal and the second is to bring the level tube parallel to the line of sight.

To make the line of sight horizontal, sight a level staff set up a short distance away and take the reading when the level bubble is central.   Then reverse the telescope to bring the level tube to the other side of the instrument, level the telescope, and again sight and read the staff.   If the instrument is in adjustment, the two readings should be the same.   If they are not the same, take the mean reading, and, using the tilting screw, set the instrument to this reading.   The line of collimation will now be horizontal, but the bubble will no longer be at the centre of its run.   Hence, the second stage of the adjustment consists in bringing the bubble central by means of the adjusting screws at the end of the level tube.   When this has been done, the line of sight and the bubble will both be horizontal, and consequently the two will be parallel.

15. Parallel-plate Micrometer.

The parallel-plate micrometer is an attachment which is sometimes fitted to precise levels to enable fine readings to be taken.   It is made up of a disc of plane glass, with its opposite faces most carefully ground to flat parallel surfaces, mounted in front of the objective in such a way that it can be tilted slightly away from the vertical and the amount of tilt measured on a graduated drum or arc.   Tilting the plate causes a horizontal ray from the staff to be bent slightly in its passage through the glass, and to emerge from the plate parallel to the original ray but displaced slightly in a vertical direction with reference to it.   Thus, in fig. 5.11, the ray **ab** from the staff is bent (refracted) at the surface of the parallel plate, and follows the straight line **bc** in the glass to emerge as the ray **cd** parallel to **ab**.   Hence, instead of sighting the point **e**, which is a point on the staff where a straight line drawn as a continuation of the ray **dc** would meet the staff, the point actually sighted is the point **a**. Consequently, in passing through the plate the ray has been displaced vertically by the amount **ae**. The glass and the tilting device are so designed that a complete revolution of the graduated drum, or the movement through a whole number of divisions of the graduated arc, causes the line of sight to be raised or lowered exactly through one or two of the smallest divisions on the staff. Hence, to use the attachment, the instrument is levelled and sighted on the staff, and the disc is tilted until the cross hair appears to coincide with one of the staff divisions, the amount of tilt

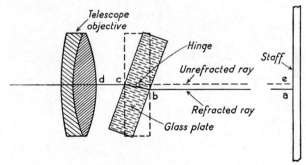

Fig. 5.11.—Parallel-plate micrometer

necessary to do this being measured, in terms of staff subdivisions, on the drum or arc. The reading is then the reading on the staff plus the reading on the micrometer drum or vernier. In this way, readings may be taken direct to 0·0001 ft. and by estimation to 0·00001 ft.

16. The Levelling Staff.

The process of levelling may be considered to be one by which vertical distances are plumbed or measured below a line in a horizontal plane, the horizontal line in this case being the line of collimation of the level, and the plane the horizontal plane through the horizontal cross hair of the instrument. The staff is the scale on which these distances are measured.

The levelling staff is a wooden rod with a flat surface on which the graduations are painted or marked, and its length when in use generally is from 6 ft. to 14 ft. The bottom of the staff is fitted with a metal shoe, usually made of bronze or gun-metal, with a flat surface underneath perpendicular to the main length of the staff. In ordinary

British and American practice, the graduations are in feet, and tenths and hundredths of a foot.   On the Continent, the graduations are almost invariably in metres, decimetres and centimetres.

There are four main kinds of staff—ordinary, telescopic, folding and target.   The *ordinary staff* is merely a single long flat piece of wood about 3 in. wide and 1 in. thick, with a metal shoe on the bottom,

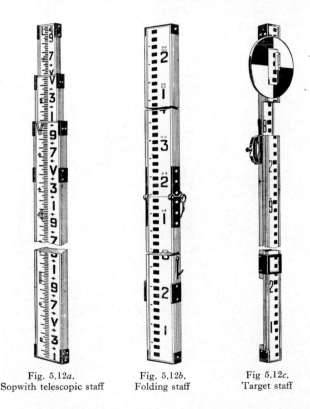

| Fig. 5.12a. | Fig. 5.12b. | Fig 5.12c. |
| Sopwith telescopic staff | Folding staff | Target staff |

and carrying the graduations on one face.   Sometimes the rod is strengthened against warping and bending by another long piece of wood, about 2 in. wide and 1 in. thick, fastened with its narrow side against the back of the staff, the whole rod then being T-shaped in section.

The *Sopwith Telescopic Staff* (fig. 5.12a) is normally made in three sections which "fold up" like a telescope.   The bottom and middle sections are in the form of a long, hollow, rectangular box open at the top, the middle section fitting into the lower one, and the upper section, which is a solid piece of wood, fitting into the middle section. Thus, when the staff is folded up it is only a little longer than the length of the lower section. When it is extended to its full length, the two upper sections are pulled out until they are engaged and held by a spring catch. In the ordinary telescopic staff, each section is about 5 ft. long, but a short staff in which each section is only about 30 in. long is made for use in mining work where head room may be much restricted.

The *Folding Staff* (fig. 5.12*b*) is hinged in the middle so that the upper half when not in use can be folded over to lie against the lower half. When fully extended, the upper part is kept in position by a pin passing through holes in plates at the sides of the hinge.

The *Target Staff* (fig. 5.12*c*) is much used in the United States and in Canada but rarely in England. The staff shown in the illustration closes to 7 ft. 4 in. and extends to 13 ft. It is graduated in feet, tenths and hundredths, and a vernier attached to the target enables readings to be taken to a thousandth part of a foot. The leveller, having levelled his instrument, signals to the staff holder to raise or lower the target until the horizontal hair of the instrument appears to intersect the middle line of the target. The staff holder then clamps the target and takes the reading, using the vernier to obtain the thousandth part. For low readings, under 7 ft., the upper part of the staff is lowered until the wooden step on it is in contact with the top of the lower part, and the target can then be slid up or down and set at the correct height. When higher readings have to be taken, the target is set at the top of the staff, and the upper part of the latter raised until the middle line of the target appears in line with the horizontal hair of the instrument. The reading is then taken on the back of the staff, where a second vernier enables readings to be taken to a thousandth of a foot.

The target staff is not so rapid in use as an ordinary plain staff, since it takes time to adjust and set the target. In addition, it has the disadvantage that the responsibility for reading the staff lies with the staff man and not with the surveyor. The advantages of it are that it enables readings to be taken direct to thousandths of a foot, but this does not necessarily mean that the real accuracy thereby obtained is much better than with an ordinary staff, because errors in bisecting the target image with the cross hairs are of much the same order as those arising from reading a plain staff to hundredths in the ordinary way. Experience, in fact, has shown that, with sights of normal length, the accuracy to be obtained from both staffs is very much the same when each is used with the same instrument. A target staff, however, is sometimes invaluable when very long sights, such as those over a very wide river, cannot be avoided.

The actual form of the graduations and figuring in different staffs shows considerable variations, those shown in fig. 5.12 being typical examples of simple straightforward types. Special forms of division and lettering have been devised from time to time to ensure easy and rapid reading and to guard against mistakes, examples of these being the Gayer staff manufactured by Messrs. Cooke, Troughton & Simms, Ltd., and the Barron levelling staff marketed by Messrs. Hilger and Watts, Ltd. Whatever form is used, the surveyor must make himself familiar with the graduations and figuring and keep throughout to a consistent system of reading. Thus, it does not matter whether he takes the top or the middle of a division as representing the value of it so long as he keeps always to the same system. At the beginning, difficulty may be experienced through the fact that the numbers appear upside down and the graduations increase in value from the top downwards when the staff is viewed in the ordinary inverting telescope. This difficulty disappears after a very little practice.

In many staffs, the graduations are on varnished strips of paper which are pasted on the face of the staff, new strips being obtainable to replace those worn out. In others, the graduations and figures are painted on the wood. For precise work it is at least desirable to use staffs in which the graduations have been cut into the surface of the wood, and for geodetic levelling, or for other levelling operations in which special accuracy is needed, *Invar staffs* are employed. This type of staff consists of a long, narrow strip of Invar (p. 31) fixed to the staff at the bottom, but let into the wood carrying the numbers in such a manner that it can move independently of the wood and be nowhere constrained by it except at its single point of attachment. In this way, the Invar is not affected by the expansion or contraction of the wood brought about by changes in temperature and humidity, while its own low coefficient of thermal expansion tends to minimize errors due to the effects of temperature changes.

It is essential that, whatever its nature, the staff should be held vertically at the point where it is being observed. For this purpose some staffs, particularly those for precise work, are fitted with a small circular level fixed at a convenient height at the rear or to one side. If no such level is fitted, the staff man should support the staff in front of his face with the palms of the two hands pressed lightly against each of the sides. When observations are being taken, he should wave the staff very gently forwards and backwards. The readings will then appear in the instrument to increase and decrease, and the reading should be taken when it reaches a minimum value.

17. Using the Level.

In levelling operations the work generally starts from a permanently fixed point whose elevation above some horizontal plane taken as a datum plane is known. Such a point is known as a *bench mark*. The staff is held on this point in a vertical position with the graduated surface facing the instrument, which is set up at some convenient distance away. When the instrument has been levelled, the reading is taken on the staff. This reading added to the known elevation of the starting point gives the height of the line of collimation above the datum plane. If rough distances are needed, readings are taken on the stadia hairs as well as on the central hair. The staff is now held over a point whose elevation is required, and the instrument, still in the same position as before, is focused on it; then, care being taken to see that the bubble is central, another reading is taken. This reading, subtracted from the height of the line of collimation which has just been determined, gives the elevation of the new point. In this way, the elevations of a number of points can, if necessary, be obtained from the one set-up of the instrument.

If other elevations are needed which necessitate the instrument being moved to a new position, a peg is driven or a good firm point selected on which to set the staff. A reading to this *turning point* is taken immediately before the instrument is moved, the elevation of the point thus being determined. The instrument is now set up and levelled at the new position, the staff held on the turning point and a reading taken. This will enable the new elevation of the line of collimation to be calculated so that, by taking readings to the staff held at different new points, the elevations of these points can easily be found.

18. Elimination of the Effects of Instrumental Errors by the Use of Special Methods of Observation.

In levelling, the effects of collimation error in the instrument can be eliminated by keeping all sights from the one station equal in length. This is seldom possible, but in running important continuous lines of levels from one point to another it is always well to try to keep the lengths of backsights (the sights necessary to obtain the height of the line of collimation) and foresights (the sights necessary to establish the heights of new turning points) from the same station as equal as possible. Stadia hairs are particularly useful for this purpose. If the instrument is of the reversible type (p. 133), collimation error can also be eliminated by taking the mean of two observations, the one with the level tube on the observer's right and the other with it on his left.

Errors due to the staff not being held vertical can be eliminated to a very large extent by using a circular level, or a plumb bob attached to the staff, or else by causing the staff to be waved gently backwards and forwards in the manner described on p. 139 and taking the smallest reading.

Readings very close to the ground should be avoided as far as possible, as they are likely to be upset through the rays from the staff having to pass through layers of air abnormally heated by direct contact with the earth.

For the rest, good levelling depends on the instrument being set up on good solid ground where it cannot move or settle between observations, and selecting good hard points, not liable to movement, for turning points, as it is obvious that any settlement of the instrument or of a turning point between observations will lead to error.

If the level has to be set up on very soft ground, three long solid pegs should be driven in such positions that the tripod legs can be set on them. At the same time it may be necessary to lay down a platform of brush or timber built around the instrument, but not touching the tripod legs or supporting pegs, to form an independent support for the observer.

THE BAROMETER, ANEROID, AND THE HYPSOMETER

19. Characteristics and Uses of the Barometer and Aneroid.

Barometers and aneroids are used in surveying for rough determinations of elevations, or differences of elevation, and a form of aneroid is also used in aeroplanes engaged on air surveys for determining the height at which the plane is flying when the photographic plates are exposed. In addition, barometers or aneroids, graduated in terms of inches or centimetres of mercury, are often needed to provide the necessary data for calculating the refraction correction in certain kinds of astronomical observations.

In ordinary surveying, the mercurial barometer is seldom used for the direct determination of elevations because of the difficulty of transporting it without damage from station to station. Often, however, and especially when a number of surveyors are working together, one is carried by the chief of the party to check the readings of individual aneroids.

Levelling with barometer or aneroid is nothing like so accurate as levelling with the surveyor's level. For this reason, these instruments are normally used only for topographical and reconnaissance work on a small scale, where great accuracy in the determination of elevations is not essential. The mercurial barometer is the more accurate instrument of the two, but the difficulty of carrying it about, and the ease with which it is broken, make it an inconvenient instrument for ordinary everyday topographical work.

20. The Mercurial Barometer.

There are two main types of mercurial barometer—*cistern* and *siphon*—and both depend on the principle of balancing a column of mercury against the pressure of the atmosphere, the atmospheric pressure at the point of observation being a function of the elevation of the point above sea-level.

In the cistern type of barometer, the mercury is contained in a glass tube about 33 in. long, the upper end of the tube being closed, and the open lower end immersed in a cistern containing mercury. The tube is exhausted of air, so that the space above the top of the mercury is a vacuum, but the surface of the cistern is open to the atmosphere. The pressure of the latter acting on the surface of the mercury in the cistern maintains a column of mercury in the tube, the upper end of this column being free from pressure of any kind, because of the vacuum, so that the height of the column depends directly upon the pressure on the surface of the mercury in the cistern.

In the *Fortin* type of cistern barometer, the cistern is made of a leather bag contained in a metal tube terminating in a cylinder of glass, through which the mercury surface can be seen. Above the surface of the mercury is a small, fixed, pointed peg to act as an index from which the height of the mercury is measured. A screw at the base enables the bottom of the bag to be raised or lowered slightly, thus raising or lowering the surface of the mercury in the cistern. When observing is taking place, the screw is turned until the surface of the mercury is seen through the glass window to be touching the point of the index pin. The height of the mercury in the tube is measured by a vernier working against a scale on a metal casing outside the glass tube, the vernier being moved by means of a knurled knob controlling a rack-and-pinion movement until the lower horizontal edge of the vernier plate appears to touch the top of the curved meniscus of the mercury column. As the vernier plate, being separated from the mercury by the thickness of the wall of the glass tube, does not actually touch the mercury surface, a mirror behind the tube is fitted to enable parallax to be eliminated. Accordingly, when the vernier is set, the bottom of the vernier plate, its image in the mirror, and the top of the mercury should all appear to be coincident. A thermometer attached to the casing enables the temperature at the time of observation to be noted. In a somewhat cheaper form of Fortin barometer the leather bag is replaced by a glass bowl which can be raised or lowered by a knurled ring to bring the mercury surface into contact with the index pin.

In the *Kew* type of cistern barometer now used for standard meteorological observations the bowl is fixed, no index point is fitted, and the scale is constructed so as to allow for the alteration in the level of the mercury in the bowl as the level of the mercury in the tube alters.

In the siphon type of mercury barometer, the lower end of the tube containing the mercury is bent in the form of a U-tube in which one branch is only a few inches long. The upper end of the short branch has a small opening through which air can be admitted, while the long branch is closed, and has a vacuum at the top in the usual way. This type of barometer in general is inferior to the cistern type, and is consequently not much used in survey work.

Mercurial barometers need to be supported vertically, and are therefore often suspended by some form of gimbal mounting attached to a special tripod. When a Fortin barometer is being transported from place to place, the adjusting screw of the bag must be tightened until the mercury is forced to fill the tube completely, and the barometer must then be carried upside down. Otherwise, if the mercury is free to move in the tube there is every chance that the tube will be shattered.

When barometrical observations are in progress, temperatures should be read on two thermometers, the one built into the instrument casing and the other held free in the air. Readings on a wet-and-dry bulb thermometer are also necessary if the most accurate results are to be obtained.

Formulæ for the conversion of the corrected pressures into heights above sea-level are available (p. 145), but in practical work the reduction is most easily made with the assistance of special tables such as those given in Close and Winterbotham's *Textbook of Topographical Surveying* (H.M. Stationery Office).

21. The Aneroid Barometer.

The operation of the aneroid barometer depends ultimately on the bulging of a thin metal disc when the two faces are exposed to different pressures. The main working part therefore consists of a circular, corrugated, air-tight box of thin metal—usually very thin German silver—which is emptied of air [see (1) in fig. 5.13]. In normal circumstances the pressure of the atmosphere would cause the box to collapse and the two faces to come together. This movement is prevented by a spring (2) supporting the upper face, the lower face being attached by a central vertical post (3) to a circular base-plate (9).

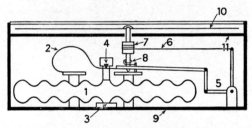

Fig. 5.13.—Diagrammatic section of an aneroid

Accordingly, the upper face of the box will move up or down according to the variations in the atmospheric pressure. The upper surface of the box carries another vertical post to which is attached a knife edge (4) bearing down on top of the spring (2), this spring being pivoted at its far end to one of a series of links (5), which transmit the motion of the spring to a light chain (6) working round a vertical spindle (7). The upper end of this spindle carries a pointer (10), and the lower end is attached to a very light hairspring (8) opposing the motion imparted by the chain. In this way, the movements of the faces of the vacuum box are transmitted to the pointer, the movements of which are read on the scale (11) on the face of the instrument.

The general external appearance of the aneroid barometer is shown in fig. 5.14 (Plate X, p 132). There are two scales, the inner one being graduated as a pressure scale in terms of inches or centimetres of mercury and the outer scale in terms of heights in feet or metres. In the more expensive instruments, the inner pressure scale is not regular, but the instrument is compensated so that the height scale is regular, thus permitting the use of a vernier for fine readings. If the instrument is of good quality, the best results will be obtained by ignoring the height scale, observing the pressure scale, taking the air temperature at the moment of observation, and converting the pressure readings into heights by means of special tables, such as those mentioned on p. 143.

In many instruments the height scale is on a ring which can be rotated relative to the pressure scale, and can be set so that the height of the station appears directly under the pointer. This is a convenient method of obtaining direct readings of altitude, but it is only useful for comparatively rough work as it introduces error by reason of the fact that differences of altitude on the height scale are not directly proportional to differences of pressure on the pressure scale. In addition, an adjusting screw at the back of the instrument allows the pointer to be moved and set to agree with the pressure registered at any time on a standard mercurial barometer.

Although the aneroid is graduated to give absolute heights above sea-level under standard conditions of atmospheric pressure, temperature and humidity, in practice these conditions never exist simultaneously in the field at any one time. Hence, it is best not to rely on readings of absolute height but to use the instrument to determine differences of height. This is done by reading the aneroid at two stations and subtracting the one reading from the other. The difference, when added algebraically to the height of the first station, will give the height of the second. Accuracy depends on atmospheric conditions remaining stable during the interval between the first and second set of readings. This seldom, or never, happens, but the effects of changes can be allowed for by a second observer taking frequent observations on another aneroid at the first station until the second can be reached, both observers booking the time of each observation. Alternatively, if the stations are not too far apart, the observer can return to the first station as soon as possible after he completes those at the other and then take another set of readings there.

The aneroid barometer can be observed either suspended vertically or laid down horizontally, but, whichever way is adopted, the instrument should always be read while it is in the same position. In all cases, the glass cover should be tapped very gently with the top of a pencil when a reading is being taken.

22. Relation between Atmospheric Pressure and Elevation of Station.

The old international formula linking atmospheric pressure with elevation of station is

Difference of elevation in feet =

$$60370(\log H_1 - \log H_2) \times \left(1 + 0{\cdot}00264 \cos 2\phi + \frac{h_1 + h_2}{R} + \frac{3P}{8H_m}\right)$$
$$\times \{1 + 0{\cdot}002036(t_m - 32)\},$$

where H_1 and H_2 are the barometer readings in inches at the lower and upper stations; h_1 and h_2 are the elevations in feet of the upper and lower stations, and R is the radius of the earth in feet at the place of observation; ϕ is the mean latitude of the two stations; P is the mean water-vapour pressure; $H_m = \frac{1}{2}(H_1 + H_2)$ and t_m is the mean temperature at the two stations in degrees Fahrenheit.

P, the humidity factor, is calculated from

$$P = P_w - 0{\cdot}00045 H_m(t_d - t_w),$$

where t_d and t_w are the readings on the dry-and-wet bulb thermometers in degrees Fahrenheit, and P_w, the saturation prsssure of water vapour, is taken from special tables such as those given in Vol. II of the Royal Geographical Society publication *Hints to Travellers*.

The terms in $\cos 2\phi$ and $(h_1 + h_2)/R$ are introduced into this formula to allow for the variation of the force of gravity with latitude and elevation of station.

The above formula is based on the assumption that the temperature of the column of air above the ground at each station remains constant at 20° C. at all elevations. It is well known that this is not so and that the temperature of the air decreases with height up to certain limits at a rate of approximately 3·566° F., or 1·981° C., per 1000 ft. This rate is known as the *lapse-rate*, and, although the graduations of the older aneroids have been based on the expression given above, in recent years this formula has given place to a new one, known as the *lapse-rate formula*, in which account is taken of the variation of the temperature of the air column with altitude. This, in its simplest form, is

$$\frac{p}{p_e} = \left(\frac{T_e - kh}{T_e}\right)^{\frac{g}{kc}},$$

where p_e is the atmospheric pressure at sea-level; p the corresponding pressure at elevation h ft. above sea-level; T_e and T are the *absolute* temperatures* at sea-level and at elevation h respectively; k is the lapse-rate in degrees Centigrade per foot rise; g is the acceleration of gravity, and $\dfrac{1}{c}$ is a physical constant, known as the *gas constant*, given by the expression

$$\frac{\rho T}{p} = \frac{1}{c},$$

where ρ is the density of the air at any given pressure p and absolute temperature T.

Assuming that at sea-level in latitude 45° a cubic metre of air weighs 1·2257 kilogrammes when $g = 980 \cdot 62$ cm./sec.², temperature = 59° F. or 15° C. (288° absolute), and the barometric pressure reduced to 0° C. is 760 mm. of mercury, this formula gives

$$p = p_e[(288 - 0 \cdot 00198h)/288]^{5 \cdot 256},$$

where p and p_e are in the same units, say inches, and h is in feet.

When temperatures differ from standard, the correction for temperature to heights obtained by this formula is given by

$$h_c - h = 0 \cdot 001929h(t - t_h)\left\{1 + 3 \cdot 44 \times 10^{-6}h + 1 \cdot 6 \times 10^{-11}h^2 + \ldots\right\}$$

in which h_c is the corrected height, h the height obtained from the formula, t the observed temperature in degrees Fahrenheit, and t_h the standard temperature in degrees Fahrenheit for height h as given by the lapse-rate formula.

$$t_h = (59° - 0 \cdot 003566h).$$

The derivation of the simple lapse-rate formula is given in the Appendix, page 197.

In the first edition of this book, the second and third terms of the formula for $h_c - h$ were not given, nor were they used in computing the temperature correction Tables in the official War Office publication *Aneroid Tables Based on a Standard Atmosphere and a Standard Lapse-rate*, which is now obsolete and out of print. More recently, however, (1957), Col. D. R. Crone, late of the Survey of India, has worked out these terms and shown that it is necessary to take them into account when accuracy is required, and, when this is done, the lapse-rate formulae give results which agree very closely with those obtained by using the international isothermal formulae given at the top of page 145.

---

* The absolute temperature is the temperature measured from *absolute zero*, or −273·16° C. In degrees absolute Centigrade it is $t + 273 \cdot 16°$, where t is the temperature recorded on the thermometer, and in degrees absolute Fahrenheit it is $t + 459 \cdot 69°$.

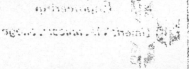

Instruments graduated in accordance with the lapse-rate formulae normally give a reading closer to the real value than those graduated on the isothermal formula. They are therefore desirable where immediate estimates of height are required as, for instance, in aircraft altimeters.

When rigorous reductions for observed temperatures and humidities are to be made, the isothermal formula admits of a very much simpler calculation of the corrections. The formula is

$$\Delta h_c - \Delta h = \frac{t_m - 68}{528} \Delta h,$$

where Δh_c is the corrected height difference;

Δh is the indicated height difference;

t_m is the mean temperature in degrees Fahrenheit at the stations of observation.

The discrepancy from the lapse-rate reduction varies as the cube of the height difference, and is only about 4ft. in an interval of 10,000 ft.

The isothermal reduction may be applied to readings made with instruments graduated by the lapse-rate formula by regarding the instrument as divided isothermally with a calibration correction varying over the range of the scale. This calibration correction is best obtained by calibration in a vacuum chamber but may be taken from standard tables or obtained by computation.

23. The Hypsometer.

The *hypsometer*, or *boiling-point thermometer*, depends for its working on the fact that the temperature at which water boils varies with the atmospheric pressure. It therefore consists of a specially sensitive thermometer, graduated to $0.2°$ F., fitting into a special vessel in such a way that its bulb is a little above the surface of water contained in a receptacle in the vessel. The water is heated by a spirit lamp and, when the water boils, the bulb of the thermometer is surrounded by steam. The temperature recorded on the thermometer is read after the water has been boiling for a couple of minutes or so, and the reading has become stationary.

The hypsometer is now little used by British or American surveyors as it is no more portable than the aneroid, nor any more accurate, and, because of the time taken in waiting for the water to boil, it is not so convenient. It is, however, used to some extent by Continental surveyors.

School of
Engineering
Limerick Technical College

The readings of the hypsometer are most conveniently converted into elevations by means of special tables, but, if need be, elevations can be computed by the formula

$$E = t(521 + 0.75t) \times \left(1 + \frac{T_m - 32}{450}\right),$$

where E = elevation in feet above point at which water boils at 212° F.

$t = 212°$ — observed temperature.

T_m = mean temperature of air at station in degrees Fahrenheit.

Water boils at 212° F. at sea-level at an atmospheric pressure of 29·921 in. of mercury and an air temperature of 32° F. In practical work, however, it is best to use the instrument to determine differences of elevation rather than to depend on calculated or tabulated absolute values of E. Suitable tables for use with the hypsometer are given in Close and Winterbotham's *Textbook of Topographical Surveying* (H.M. Stationery Office), and in Vol. I of the Royal Geographical Society publication *Hints for Travellers*.

As it is only possible to estimate readings on the thermometer to about 0·1° F., and since a difference of 0·1° F. in the reading of the thermometer corresponds to a difference in elevation of about 50 ft., it will be seen that this fact alone sets a severe limitation on the possible accuracy of work with the hypsometer.

CHAPTER VI

OPTICAL MEASUREMENT OF DISTANCE

1. Principal Methods.

There are several methods of measuring distances by optical means, but the three most commonly used are:

1. Stadia methods, or tacheometry (tachymetry).
2. Subtense methods.
3. Rangefinder.

All three depend on the fact that if the length of the base and the apical angle of an isosceles or right-angled triangle are known, the length of the perpendicular from the apex, which is the quantity required, can easily be calculated. In the first method the apical angle is fixed in the instrument and the base is measured: in the second, the base is fixed but is not part of the instrument and the apical angle is read on the latter: in the third method, the base is fixed in the instrument, and the required distance, which is the quantity that is read directly, depends on what is in effect an observation of the apical angle.

Of these methods, tacheometry in some form or another is much more extensively used in engineering surveys than either of the other two, both of which are used mainly in topographical or military work.

TACHEOMETRY

2. Theory of Tacheometry.

In fig. 4.16c, d and e (p. 83) we have shown a diaphragm with two extra horizontal hairs equally spaced on either side of the central hair. These additional hairs are called *stadia hairs*, and the diaphragms of most theodolites and levels are now fitted with them. In fig. 6.1 (p. 150), let **a** and **b** be these hairs, O the optical centre of the object glass of the telescope, and ST a levelling staff held at distance v from O. Let u be the distance of the stadia hairs from O and PQ be the vertical axis of the instrument, distant c from O and d from ST. Let $\mathbf{x}$, $\mathbf{y}$ be the points on the staff whose images are formed at $\mathbf{a}$ and $\mathbf{b}$ respectively. Then the length $\mathbf{xy} = s$ is the *staff intercept* obtained by subtracting the reading of hair $\mathbf{a}$ on the staff from the reading of hair $\mathbf{b}$. Let i be the length of $\mathbf{ab}$ and f the focal length of the lens.

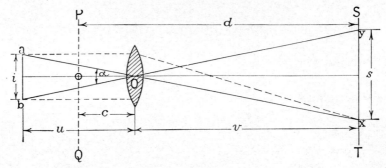

Fig. 6.1.—Diagram illustrating the principle of tacheometry

We have from the figure

$$d = v + c,$$

and

$$\frac{i}{s} = \frac{u}{v}.$$

Also, from the formula for the formation of an image by a lens,

$$\frac{1}{v} + \frac{1}{u} = \frac{1}{f}.$$

$$\therefore \ \frac{1}{v} + \frac{s}{iv} = \frac{1}{f}.$$

$$\therefore \ v = \frac{(i+s)f}{i} = f + \frac{s}{i}f.$$

$$\therefore \ d = \frac{s}{i}f + (f + c).$$

Now f and i are constants for the instrument and hence we can write $k_1 = f/i$, where k_1 is a constant known as the *stadia constant*. Consequently,

$$d = k_1 s + (f + c). \tag{1}$$

Thus, the distance from the staff to the instrument is k_1 times the staff intercept plus $(f + c)$. The quantity c is not quite a constant because it varies as the instrument is focused, but the variation due to focusing is small in comparison with the other quantities or with the errors inherent in the method. Hence, for all practical purposes we can regard $(f + c)$ as a constant of the instrument which we shall call k_2. Thus, we have

$$d = k_1 s + k_2. \tag{2}$$

Usually the stadia hairs are set so that the value of k_1 is very approximately 100, and we then have the simple equation

$$d = 100 s + k_2. \tag{3}$$

In general, however, we cannot always rely on the constant k_1 being exactly equal to the value given by the makers, and it is therefore well to determine both k_1 and k_2 by observation. This can easily be done by taping out several distances and observing the staff intercept at each distance. From each pair of observations we get a couple of simultaneous equations of the form

$$d_1 = k_1 s_1 + k_2,$$
$$d_2 = k_1 s_2 + k_2,$$

where d_1, d_2, s_1, s_2 are observed values of d and s, and, from these, values for k_1 and k_2 can be derived from different pairs of observations and mean values accepted. Better still, if the surveyor is familiar with the method of least squares, a series of observation equations can be formed, and the " most probable " values of k_1 and k_2 found from them.

In practice, there are many types of work in which the constant k_2, which is generally of the order of somewhere about a foot to a foot and a half, is neglected, but, once the value of the constant is known, it is not very difficult to take it into account.

The above theory assumes that the instrument is level, with the vertical axis vertical and the line of collimation horizontal, and that the telescope is of the external focusing type. If an internal focusing telescope is used, matters are complicated by the fact that the focus of the objective system is variable and not fixed. This variation is small for sights over about 100 ft., but for short sights it causes an appreciable variation in the value of k_1. For most practical purposes, however, formula (2) can be used for anything but very short sights.

3. Tacheometry on Slopes.

When a theodolite is used, the line of collimation is generally inclined to the horizontal and the simple formula needs modification. In this class of work the difference of height between the staff and the instrument is required as well as the horizontal, as opposed to the inclined, distance between them. In addition, two cases arise:

1. When the staff is held vertical.
2. When the staff is held perpendicular to the line of collimation.

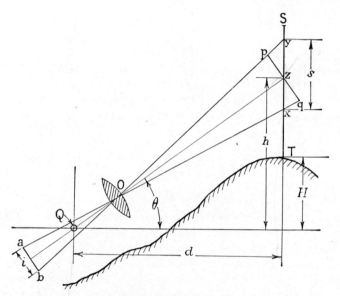

Fig. 6.2.—Tacheometry on slopes—vertical staff

In fig. 6.2 we take the case of the staff held vertically on a point T, height H above Q, the horizontal axis of the instrument. The line of collimation O**z** meets the staff at **z** and the lines of sight through the stadia hairs meet it at the points **x** and **y**. Consequently, $\mathbf{xy} = s$ is the stadia intercept obtained by subtracting the reading of the upper stadia hair from the reading of the lower one. Let θ be the inclination of the line of collimation to the horizontal.

In applying formula (2) above we cannot use s directly in the ordinary way because here it is inclined to the line of collimation. Through z draw **qzp** perpendicular to O**z** to meet O**x** and O**y** in **q** and **p**. Then, by formula (2),

$$Qz = k_1 pq + k_2.$$

But **pOq** is a very small angle, and we can consider the angles **zqx** and **zpy** to be right angles. Hence,

$$pq = yx \cos \theta = s \cos \theta.$$

$$\therefore \ Qz = k_1 s \cos \theta + k_2.$$

Also, we see that

$$d = Qz \cos \theta$$

$$= k_1 s \cos^2 \theta + k_2 \cos \theta, \tag{4}$$

$$h = Qz \sin \theta$$

$$= k_1 s \cos \theta \sin \theta + k_2 \sin \theta, \tag{5}$$

and

$$H = k_1 s \cos \theta \sin \theta + k_2 \sin \theta - R \tag{6}$$

$$= \tfrac{1}{2} k_1 s \sin 2\theta + k_2 \sin \theta - R, \tag{6a}$$

where R is the reading **Tz** of the central hair on the staff.

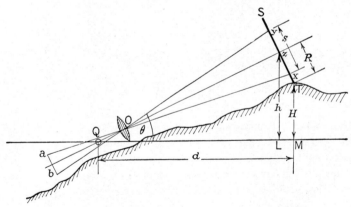

Fig. 6.3.—Tacheometry on slopes—inclined staff

If the staff is held perpendicular to the line of collimation as in fig. 6.3, we have from (2),

$$Qz = k_1 s + k_2.$$

But,

$$d = QL + LM,$$

where L and M are the feet of the perpendiculars from $\mathbf{z}$ and T on a horizontal line QM through Q, and

$$QL = Q\mathbf{z}\cos\theta,$$

$$LM = R\sin\theta.$$

$$\therefore\; d = k_1 s\cos\theta + k_2\cos\theta + R\sin\theta. \tag{7}$$

Also,

$$H = h - R\cos\theta$$

$$= k_1 s\sin\theta + k_2\sin\theta - R\cos\theta. \tag{8}$$

When the staff is to be held perpendicular to the line of collimation, a small tube fitted with cross hairs to provide a line of sight at right angles to the staff is generally attached to the latter. The staff is then inclined until the cross hairs appear to intersect the horizontal axis of the theodolite. This sighting tube is normally at the ordinary height of a man's eyes above the bottom of the staff, although theoretically it should be at the point where the line of collimation of the theodolite meets the staff.

For staffs held vertically, a plumb bob attachment or a small circular spirit level to show when the staff is vertical is almost a necessity.

The principal formulæ in tacheometry are (2), (4), (6), (7) and (8), and it is well to commit these to memory, although they can always be deduced very simply.

4. Anallactic Lens.

The additive constant $(f + c)$ in formula (1), and its appearance in the formulæ relating to observations of sloping lines, adds terms in each formula which it would be well to avoid if this were possible, and the question arises, could this be done in some way by, say, the introduction of a supplementary lens? In our original diagram, fig. 6.1 (p. 150), we note that $v = \tfrac{1}{2}s\cot\tfrac{1}{2}\alpha$, where $\alpha = 2\tan^{-1}\dfrac{i}{2u}$ is the apical angle $\mathbf{x}O\mathbf{y}$, so that in this case v, but not d, is directly dependent on s and can be expressed in the form $v = k_1 s$, where k_1 is the stadia constant. Hence, this suggests fig. 6.4b, in which we can see that, if we could replace the object glass by an optical system which would be equivalent to an objective O' placed on the vertical axis of the instrument at distance c from O, we would have $d = \tfrac{1}{2}s\cot\tfrac{1}{2}\beta = ks$, where $\beta = 2 \tan^{-1} \dfrac{i'}{2u'}$, i' being the image of **xy** formed by O' and u' the distance of this image from O', and so d would be determined directly in terms of s without the addition of any $(f + c)$ term. It can be shown that this is possible if a suitable and suitably placed supplementary plano-convex lens is introduced between O and the

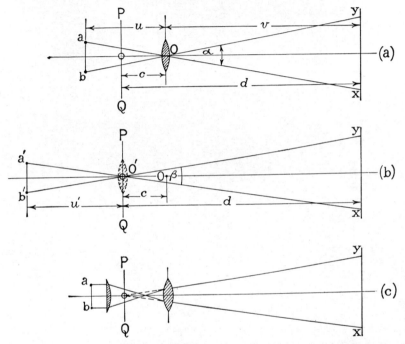

Fig. 6.4.—Diagram illustrating the principle of an anallactic lens

diaphragm as shown in fig. 6.4c. This lens, with the objective, then forms an optical system equivalent to the lens O' in fig. 6.4b.

Fig. 6.5 shows the passage of the rays through the supplementary lens L and the objective O. Parallel rays from the stadia hairs **a** and **b** are brought to a focus at F, where F is the focus and LF is the focal length f_1 of the lens L, and help to form a virtual image **a'b'** at distance u from O. The rays after passing through F are refracted through O in such a way that, when produced backwards from O, they form a pencil diverging from O', the intersection of the vertical and horizontal axes of the instrument. In this case it can be shown that, if l is the distance between L and O,

$$l = f_1 + \frac{fc}{(f + c)} \tag{9}$$

and we have the simple relation

$$d = ks, \tag{10}$$

where the constant k is given by

$$k = \frac{ff_1}{i(f + f_1 - l)}. \tag{11}$$

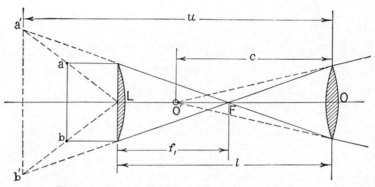

Fig. 6.5.—Diagram illustrating the rays in an anallactic lens

The supplementary lens L is called an *anallactic lens*, and its introduction is due to an Italian geodesist, Ignazio Porro, somewhere about 1830. Accordingly, it is sometimes called a *Porro lens*.

It should be noted here that, whereas the ordinary internal focusing lens moves with respect to the objective, and the effective focal length of the combination is thus variable, the distance between the anallactic lens and the objective is fixed. In theory, a truly anallactic lens, in the form of a *single* simple lens only, cannot be fitted to an internal focusing telescope, although Messrs. Cooke, Troughton & Simms have produced an internal focusing telescope which has a single supplementary lens and is practically anallactic. In addition, they have designed an internal focusing telescope which is perfectly anallactic, but in this case there are three supplementary lenses instead of one between object glass and diaphragm.

It should also be noted that the disadvantage of an anallactic lens is the loss of light through the introduction of the extra lens. This loss of light can only be countered by increasing the aperture of the objective.

5. Movable Cross Hairs.

A variation on the ordinary fixed stadia hairs and a variable staff reading is to have movable hairs on the instrument and a fixed staff intercept. In this case equation (1) may be written in the form

$$d = r K + (f + c),$$

where K is a constant proportional to fs and r a measured quantity inversely proportional to a variable value of i. In this method, which is not much used, the hairs are set to coincide with fixed marks on a staff and the reading r is read direct on a micrometer.

The method is really a form of the subtense method using the stadia principle.

6. Tacheometers.

Any theodolite fitted with ordinary stadia hairs can be used as a tacheometer, but the term *tacheometer* is usually confined either to theodolites fitted with stadia hairs and an anallactic lens, or to other instruments which also embody special graduated arcs or other fittings designed to simplify the reduction of observations of sloping lines.

In theodolites specially intended for stadia observations, the main points to be considered are high magnification, and large effective aperture of the telescope. In addition, a good sensitive bubble should be fitted on the vernier frame.

There are many special instruments fitted with devices for enabling horizontal distances and differences of elevation to be read directly on the instrument, or obtained with a minimum of computation from the instrumental readings. These include the Beaman Stadia Arc and various kinds of so-called self-reducing tacheometers.

7. The Beaman Stadia Arc.

This specially graduated arc is very commonly used on plane-table telescopic alidades as well as on ordinary tacheometers. The graduations are at the points where $\sin \theta \cos \theta$ or $\frac{1}{2} \sin 2\theta$ is equal to $0 \cdot 01$, $0 \cdot 02$, $0 \cdot 03$, &c., corresponding to values of θ equal to $0° 34' 23''$, $1° 08'$ $46''$, $1° 43' 12''$, &c. Hence, as the telescope is anallactic, and the numbering of the graduations consists of the actual values of $\frac{1}{2}k \sin 2\theta$, the difference of elevation is found, by equation (5), from

$$h = rs,$$

where r is the reading on the arc.

In order to avoid possible errors in the signs of vertical angles, the zero reading on the vertical scale is sometimes marked 50, so that in this case 50 must be *subtracted from* the scale reading, positive numbers indicating increases in elevation and negative numbers decreases in elevation.

A second scale is provided for the reduction to the horizontal of the inclined distances. This is not in terms of $k \cos^2 \theta$, because the values of this quantity are relatively large numbers which vary slowly with small values of θ, but in terms of $k(1 - \cos^2 \theta) = k \sin^2 \theta$, as this is a much smaller quantity in which the variations are numerically of much the same order as itself. Hence, in finding the horizontal distance, the staff intercept must be multiplied by the reading on the horizontal scale, and the result subtracted from the staff intercept multiplied by k, the stadia constant. In practically all cases, the stadia hairs are set so that this constant has the value 100, and the numbering of the graduations on the horizontal and vertical scales is based on this value.

From their nature, the scale divisions on the Beaman Arc are not equal in length, a fact which makes it impossible to use verniers or micrometers with them. Accordingly, after the instrument is levelled, the telescope is directed to a point approximately at the centre of the staff, and the instrument is then clamped. The fine-motion tangential screw is then used to bring the nearest graduation on the vertical arc against the index mark, after which the readings on the staff of the central and stadia hairs, together with the readings on the two arcs, are observed and noted.

8. Self-reducing Tacheometers.

These are instruments which enable the reduced horizontal distance, and in some cases vertical differences of elevation as well, to be obtained by direct multiplication of the staff intercept by the instrumental constant. Early instruments of this kind are the Stanley Compensating Diaphragm and the Jeffcott Direct-Reading Tacheometer, the first of which gives direct readings of the reduced distance and the second direct readings of both distance and difference of elevation. In both instruments the stadia wires are replaced by movable metal pointers, the vertical distance between the pointers being varied automatically in such a way that the staff intercept multiplied by a constant gives the desired horizontal distance or difference in elevation directly. Much the same idea is now embodied in certain self-reducing tacheometers manufactured by some Continental manufacturers. In these, three to four curves corresponding to the stadia wires are etched on rotatable glass discs which are seen lying in the plane of the diaphragm. As the telescope is raised or lowered about its horizontal axis, these discs are rotated by suitably designed cams and levers in such a manner that, when the staff appears to coincide with the vertical hair, the staff intercept between two of the curves, multiplied by the instrumental constant, gives the reduced horizontal distance, while the staff intercept between another two curves multiplied by a constant gives the true difference of elevation.

9. Tacheometry with Horizontally Supported Staffs.

It is obvious that stadia measurements would be possible with an instrument fitted with vertical stadia hairs, used in conjunction with a staff supported in a horizontal position, and that, with such an arrangement, the horizontal and vertical components of the inclined distance would become $(k_1 s + k_2) \cos \theta$ and $(k_1 s + k_2) \sin \theta$ respectively, expressions which are a little simpler than any of the equations (4), (6), (7) or (8). With this method a special support to hold the staff is necessary and special tripods and diaphragms with vertical stadia hairs are obtainable. The necessity for carrying extra equipment, and the general inconvenience in handling it in the field when compared with the ease of handling the simpler equipment needed for ordinary stadia work, militate against the extensive use of the method. Horizontal tacheometry, however, has one great advantage over vertical tacheometry as regards accuracy. This is because it is not subject in the same way to errors in reading due to differential refraction, i.e. to unequal bending of the rays from the ends of the intercept due to temperature variations causing unequal density in the layers of air near the earth's surface. For this reason, it would appear that horizontal tacheometry must always be more accurate than vertical tacheometry and used when the utmost precision possible is desired.

In recent years instruments known as Wedge Telemeters and Split-Image Tacheometers, which are used in conjunction with a specially-graduated horizontal staff and some of which are self-reducing, have been designed to give a much greater accuracy than that obtainable with an ordinary tacheometer. A description of these is given on pages 189-195.

10. The Tangential System.

It is possible to obtain distances from readings on a staff without the use of stadia hairs. In fig. 6.6, ST is a staff held at T and O is the theodolite. The latter, after having been levelled, is directed to a point A near the bottom of the staff, and the staff reading and the angle of inclination ϕ of the line of collimation observed. Similar observations are taken to the point B near the top of the staff, the angle of inclination of the line of collimation being θ. Drawing the

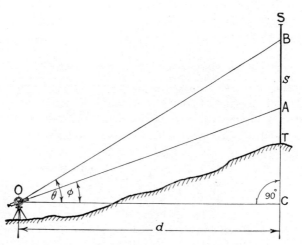

Fig. 6.6.—Diagram illustrating the tangential system

perpendicular through ST to meet a line drawn horizontally through O at C, we see that $BC = d \tan\theta$ and $AC = d \tan\phi$. Hence, if s is the difference in reading of the staff, or the staff intercept, we have,

$$s = AB = TB - TA = CB - CA$$

$$= d (\tan\theta - \tan\phi),$$

or,

$$d = \frac{s}{(\tan\theta - \tan\phi)} \text{ for angles of elevation.}$$

Similarly,

$$d = \frac{s}{(\tan\phi - \tan\theta)} \text{ for angles of depression.}$$

The disadvantage of this method is that two sights of the instrument are required in the field and, unless the instrument is set to read predetermined angles whose tangents are known, two natural tangents have to be looked out for the computation. Various types of special instruments have been devised to simplify the field work, or the work involved in the computations. Among these may be mentioned Eckhold's Omnimeter, the Bell-Elliot Tangent-reading Tacheometer, the Gradienter, and the Szepessy Direct-reading Tacheometer. The method, however, is not so commonly used as the ordinary one, and, in general, is not quite so accurate.

11. Staffs for Tacheometrical Work.

In tacheometrical work the sights are often much longer than they are in ordinary levelling, but ordinary levelling staffs are commonly employed for such work. Clear, bold graduations are essential if full advantage is to be taken of the method, and most makers provide special staffs with special forms of graduations designed to be seen clearly at a distance of anything up to, say, 500 to 800 ft.

If the staff is to be held vertically, it is desirable that it should be provided with a plumb bob or small level for ensuring that it is held truly vertical, and, if it is to be used perpendicular to the line of collimation, it should be provided with the sighting tube referred to on p. 154.

12. Accuracy of Tacheometrical Observations.

The accuracy of tacheometrical work varies with the observer and also with the conditions of observing prevailing at the time, the atmospheric conditions most unfavourable to accuracy being large variations in the temperature of the air near the ground and high wind which prevents the staff from being held steady. In normal conditions, however, experience indicates that, if very long sights are avoided, the error to be expected in distance measurements may be taken to be of the order of about 1/500 and in differences in height of about 0·2 to 0·3 ft. for ordinary angles of inclination. This is for a single sight. For long lines, where distance and elevation are carried forward by a number of set-ups, the error will tend to be proportional to the square root of the number of set-ups. The smallest fractional errors for single set-ups will be obtained with sights of ordinary length, say between 350 and 700 ft. With the split-image type of tacheometer described on pages 192-195 an accuracy of from 1/3000 to 1/10,000 in distance is obtainable with sights up to about 450 ft.

The chief sources of error in tacheometrical work are:

1. Errors in staff readings.
2. Errors in the assumed value of the stadia constant and others inherent in the instrument or in holding the staff.
3. Errors due to natural causes such as abnormal atmospheric shimmer and wind.

SUBTENSE METHODS

13. Principles and Use of Subtense Bar.

In the ordinary subtense method of determining distance, as most commonly used, a bar or rod of fixed length between signals or marks carried at the ends is supported horizontally and the horizontal angles subtended at the instrument station by the signals is measured on

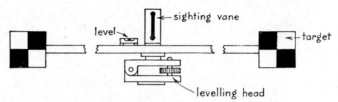

Fig. 6.7.—Diagram of a subtense bar

the theodolite. The formula for reduction is $d = \frac{1}{2}s \cot \frac{1}{2}\alpha$, where d is the distance required, s the distance between signals, and α the angle measured on the theodolite. In practice, since α is small, it is sufficient to use the approximate form $d = s \cot \alpha$ or $d = \dfrac{s}{\alpha'' \sin 1''} = \dfrac{206625}{\alpha''} \cdot s$, where α'' is the value of α expressed in seconds of arc.

The subtense bar, fig. 6.7, is usually from 3 ft. to 12 ft. long, and is mounted on a tripod through some form of levelling head which permits of easy levelling and centring and of orienting the bar at right angles to the line of sight. For the latter purpose, a sighting vane or telescope is fixed at right angles to the bar, so that when the cross hairs are directed on the theodolite the bar is at right angles to the line joining it to the instrument. A small level mounted on the bar indicates when the latter is horizontal.

The small angle subtended at the instrument is best measured by means of a horizontal eyepiece micrometer fitted in the theodolite telescope, but, if such a fitting is not available, the angle should be observed on the horizontal circle by the method of repetitions (p. 111).

Here, however, it should be noted that, if the angle is measured by micrometer, the computed distance will be the inclined distance when the sight is taken along a slope: if the angle is measured on the horizontal circle, the computed distance will be the horizontal distance, and hence for such readings no correction for reduction to the horizontal is required.

14. The " Hunter Short Base ".

During the last twenty years the Survey of India has used a special apparatus devised by Dr. J. De Graaff-Hunter for accurate subtense work which is known as the *Hunter Short Base* or *H.S.B.* This consists of a long Invar tape, 160 metres long, suspended through 12-lb. weights in two equal spans of 80 metres each between two tripods or *bipods*, and with sighting vanes set accurately over the central support and the end marks of the tape. The tape is set out at right angles to the length to be measured, and the angles subtended by each span measured with a theodolite. An accuracy of something like 1/50,000 of the distance measured is attainable with this apparatus, but its length, while contributing to accuracy in normal circumstances, make its use practically prohibitive in dense bush country owing to the amount of clearing that would be necessary to make the sighting vanes visible from the instrument station.

THE RANGE-FINDER

15. The Scope and Theory of the Range-finder.

The range-finder is used in military and also sometimes in ordinary topographical work, particularly in such operations as traversing along a wide river. In general, however, it is not much used in ordinary engineering survey work. Very much longer sights can be taken with it than are possible with stadia hairs, but, while range-finding is almost as accurate as tacheometry for short lines, accuracy falls off quickly as the length of the line increases much beyond about 500 yd., when the error becomes roughly proportional to the square of the length of the line. Thus, with a range-finder whose base is 1 metre long, the average error at 500 yd. may be taken at about 1 yd. or 1/500; at 2000 yd. about 15 yd. or 1/130, and at 8000 yd. about 240 yd. or 1/33. With a shorter instrument, the average error will be somewhat greater.

7   (G 446)

Fig. 6.8a shows two mirrors, A and B, the planes of which are perpendicular to the plane of the paper, at the ends of a line AB of fixed and known length, this length, in surveying range-finders, being usually about 3 to 6 ft. The mirror at A is fixed with its reflecting surface making an angle of 45° with AB, but the mirror at B is mounted to rotate about an axis in its plane and through its centre at the point B. Two right-angled reflecting prisms P_1 and P_2 are fixed at the centre of AB in the positions shown, with the prism P_1 mounted immediately above the prism P_2 and with the parallel plane faces of both prisms perpendicular to AB and to the plane of the paper. A telescopic objective

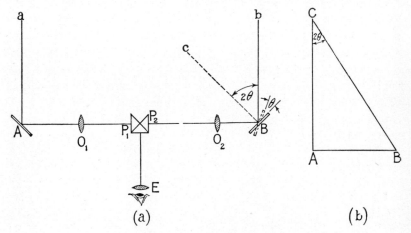

Fig. 6.8.—Diagram illustrating the principle of the range-finder

O_1 is placed between A and P_1, and a similar objective O_2 is placed between P_2 and B at the same distance as O_1 from the centre of AB. E is an eyepiece placed at a suitable distance from P_1 and P_2 to receive rays reflected by P_1 and P_2 in a direction at right angles to AB.

Consider a ray aA falling on the mirror at A and making an angle of 45° with the surface of the mirror. By the laws of reflection of light, this ray will be reflected at A to pass along the direction AB at right angles to aA. At P_1 it will be reflected at right angles to AB to pass through the eyepiece at E. Similarly, if the mirror at B is set to make an angle of 45° with BA, a ray bB perpendicular to BA at B will be reflected along BA to P_2, and thence at right angles to BA to pass through E. It will therefore be seen that the image formed in the eyepiece E by an object on bB will appear to lie immediately below the image of an object on aA. As a corollary, since the rays from a very distant object falling on the ends of AB can be considered to be parallel, the two images of a very distant object will appear to lie one above the other when the mirrors are in their normal position making an angle of 45° with AB.

Now suppose the mirror B to be turned through an angle θ in the direction shown dotted, and let a ray cB from an object C situated on the ray aA be reflected at B to pass along the line BA and then at right angles to BA at P_2 to pass through E. Two images of the same object will be seen in the eyepiece, the one image immediately above the other, and it can be shown that, by the laws of reflection, angle cBb $= 2\theta$. Consequently, by measuring θ, the angle through which the mirror has been rotated from its normal position at 45° to BA, the angle cBb can be deduced. We now have the right-angled triangle represented in fig. 6.8b, where AB represents the known distance between the mirrors, the angle at A is a right angle and the angle ACB $=$ cBb in fig. 6.8$a = 2\theta$. As we now know the angle ACB, and the length AB is also known, the distance AC can be calculated.

16. The Instrument and its Use.

In the actual instrument, scale graduations are such that the rotation of the mirror is not read directly in angular measure but in terms of the distance AC, so that the required distances are read directly and no computation work is necessary. In addition, the two images are often brought together into the same plane to overlap one another, not one above the other as described above, and the observation consists in bringing them into exact coincidence.

The method of using the range-finder will now easily be understood. The instrument, having been levelled by means of the foot screws and level bubble provided for the purpose, is rotated on its stand until an image of the object whose distance is required appears in the eyepiece. The mirror at B is then rotated until a second image of the object is seen, and appears to lie either directly below or in exact coincidence with the first image. The required distance is then read directly on the graduated scale or drum.

CHAPTER VII

INSTRUMENTS FOR SURVEYING BY GRAPHICAL MEANS

THE PLANE-TABLE

1. Plane-tabling.

Plane-tabling is a graphical method of survey in which the map is rough-drawn in the field as the survey proceeds, and is not plotted afterwards from measurements recorded in books. The method thus has the advantage that the plotted work is continually under the eyes of the surveyor while he is still on the ground. It is a method which is best suited for small- or medium-scale topographical mapping in open, or fairly open, country, where previously fixed control points, suitably marked if necessary by signals which can be seen from a distance, already exist.

2. The Plane-table.

The plane-table, fig. 7.1 (Plate XI), consists of a flat board of hard seasoned wood, usually teak, measuring 24 in. by 18 in. to 30 in. by 24 in., and provided on its under-side with an aluminium *racer ring* in the centre, aluminium battens, allowing for expansion and contraction of the wood, parallel to each of the two shorter edges, and corner fittings to prevent injury to the corners. In the centre of the racer ring there is a threaded hole to take a screw passing through the head of the aluminium casting, which forms the head of the polished teak tripod. The screw is provided at its lower end with a butterfly nut to enable the table to be attached very firmly to the tripod, and to be set and clamped in any desired direction relative to the tripod.

In the Johnson Head form of tripod head, shown in fig. 7.2 (Plate XII, p. 168), a quick-levelling adjustment, combined with an independent horizontal motion, is provided instead of the ordinary simple racer ring and tripod head. This is especially useful in large-scale work where it is important that the table should be level, with the exact point on the plan representing the point on the ground set accurately over the ground mark.

PLATE XI

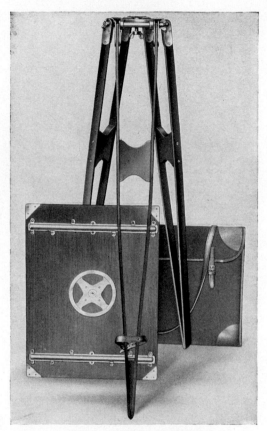

Fig. 7.1.—THE PLANE-TABLE
(By courtesy of Messrs. Hilger and Watts, Ltd.)

The plane-table is almost invariably provided with a canvas water-proof case in which it can be carried in wet weather. Such a case is necessary to protect the paper.

For sighting purposes an *alidade* of some kind is employed. Fig. 7.3 (Plate XII), shows the simple form used with the military plane-table. It consists of a polished boxwood rule with suitable scales engraved on the bevelled edges, and two metal sights, hinged to fold down on the rule when the alidade is not in use. With this alidade, the signal to be viewed is seen through slits in the sights, one slit being provided with a central vertical sighting wire, and the direction of the signal can then be drawn in on the paper by a sharp pencil laid against the edge of the rule. The string joining the two sighting vanes is sometimes used as a sighting wire when sights of considerable inclination have to be taken.

When the plane-table is used in topographical work, some means of obtaining differences of height between the observed points and the plane-table is needed. Ordinarily, if, as in the case of the simple alidade just described, the alidade is not provided with means for observing vertical angles, an *Indian clinometer*, fig. 7.4 (Plate XIII, p. 172), is used. This consists of a peep-hole carried on a short vertical arm, which is hinged at its lower end to one end of a fairly heavy, long, narrow, metal stand, a long vertical arm, with a central vertical slit, being hinged to the other end of the stand. A slide, provided with a small window, in the middle of which is a horizontal wire, can be moved up and down the longer vane by a rack and pinion fitted with a milled head. The stand is fitted with two feet at its forward end, and a tilting levelling screw at the other end, and it also carries a level tube for indicating when the instrument is properly levelled. The window and cross hair in the slide can be seen from the peep-hole in the short vane through the slit in the long vane, the latter having a scale of degrees on its inside face at one side of the slot and a scale of natural tangents at the other. In this way, sights to distant objects can be taken and the angle of inclination, or its tangent, read on one of the scales. Usually, it is the tangent scale which is read, as this reading, multiplied by the distance of the object as scaled from the map, gives the difference of elevation between object and plane-table.

When the instrument is not in use, the vanes fold down over the stand and the whole apparatus is carried in a canvas or leather case. When it is to be used, the instrument is laid on the plane-table and levelled by the levelling screw after the vanes have been lifted to the upright position, and the instrument sighted at the object whose elevation is required.

In a slightly simpler form of Indian clinometer, there is no sliding cross hair on the front vane, and the graduation on the scale which appears to be in line with the object as viewed through the slit from the peep-hole is read direct.

A small spirit level and a compass, if not indispensable, are useful additions to the ordinary plane-table equipment, the former for ascertaining if the table is properly level and the latter for orienting it to magnetic north. The level may either be of the tubular variety or of the circular type, but, whichever kind is used, it must have a flat base which can be laid on the table and is truly level when the bubble is central.

The compass used with a plane-table is a trough compass in which the longer sides of the trough are parallel and flat, so that either side can be used as a ruler or laid down to coincide with a straight line drawn on the paper.

Instead of the ordinary alidade and Indian clinometer already described, a telescopic alidade may be employed. This consists of a telescope fitted with cross and stadia hairs, vertical graduated arc and level tube, mounted on a heavy metal base or *blade* which rests on the plane-table top and can be used as a ruler. In fig. 7.5 (Plate XIII, p. 172), the blade is shown provided with a parallel-rule attachment and a circular level for rough levelling. The vertical arc is usually graduated in degrees and can be read to minutes by means of an attached vernier. If desired, a *Beaman Stadia Arc* (p. 157) can be obtained as an extra to most telescopic alidades to replace, or as an addition to, the ordinary arc. This enables the observer to obtain very quickly and easily from the stadia readings the true horizontal distance from the plane-table to a levelling staff and the difference in elevation between them.

In the Watts Microptic Alidade by Messrs. Hilger and Watts, Ltd., shown in fig. 7.5 (Plate XIII), the vertical circle is of glass, and is provided with an optical system reading directly to 10' and by estimation to 1'. A Beaman Arc scale is also fitted, and the scales on it and on the degree circle of the vertical arc are viewed simultaneously in the special eyepiece shown above the telescope. The folding mirror above the vertical circle level tube enables a complete observation to be taken from one position of the observer.

Another fitting which is practically indispensable on all large-scale work where accurate centring over a mark is desirable is a *plumbing bar*, which can be used to indicate when a point on the plan is vertically over the corresponding point on the ground. This is described on p. 172.

PLATE XII

Fig. 7.2.—JOHNSON TRIPOD HEAD
(By courtesy of Messrs. Hilger and Watts, Ltd.)

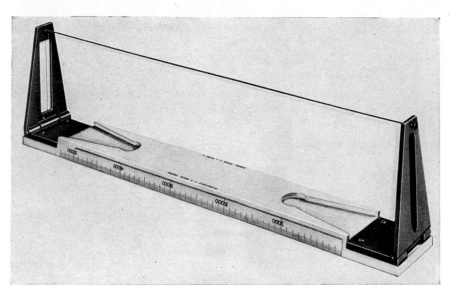

Fig. 7.3.—SIMPLE ALIDADE
(By courtesy of Messrs. Hilger and Watts, Ltd.)

3. Preparing the Plane-table.

Before work is commenced, the paper must be properly mounted on the plane-table. The paper should be best-quality Whatman mounted on linen, and should be cut with 6-in. overlaps to the shape shown in fig. 7.6, in which the shaded portions are to be cut away and the white parts left. Both sides of it should then be damped with a wet sponge and the sheet laid on a clean flat surface, paper side down. The board should then be laid face downwards on the wet linen surface,

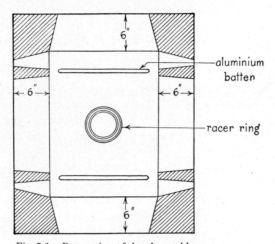

Fig. 7.6.—Preparation of the plane-table

and the overlaps at the sides bent over the edges of the board and stuck down to the under-side of the latter with good cornflour paste, the paper being stretched evenly while this is being done. If necessary, drawing pins may be used to hold the flaps down until the paste is dry. When the board has been left for about twelve hours in a cool dry place, the paper should be firmly and smoothly stretched over the surface of the board and fit for use.

Having mounted the paper, the next step is to plot on it all fixed points, trigonometrical or traverse, as well as bench marks or other points of known elevation, which come within the area covered by the map. This must be done as accurately as possible, either by means of rectangular axes, if the co-ordinates are given in terms of rectangular co-ordinates, or (as is usual in small-scale mapping) from a mesh, or *graticule*, of lines of latitude and longitude if the co-ordinates are given in terms of latitude and longitude.

When the map is completed, the paper can be cut away along the edges and detached, the flaps of paper still sticking to the under-side of the board being removed by damping.

4. Tests and Adjustments of the Plane-table.

1. *Test of Table Surface.*—The surface of the table should be a perfect plane. Test by applying a straight edge in different directions. If the surface is not true, high spots should be planed or sand-papered down.

2. *Perpendicularity of Board to Vertical Axis.*—The surface of the board or table should be perpendicular to the vertical axis of rotation. Test by putting a spirit level on the table and level the latter to bring the bubble to the centre of its run. Turn the table about its vertical axis through 180° and see if the bubble remains central. If not, correct *half* the apparent error by putting packing between the under-side of the board and the tracer ring, or else by filing down one side of the top of the latter. Re-level the table and then repeat the test and adjustment in a position at right angles to the first. Keep repeating test in both positions until the bubble is central in all directions when it is reversed.

3. *Straightness of Edge of Alidade Ruler.*—Using the ruling edge of the alidade, draw a straight line on a flat surface. Reverse the alidade end for end and put the two ends against the ends of the line. Draw a straight line between them. If this line does not coincide with the one previously drawn, the alidade is not straight. Correct by careful filing and then repeat the test.

4. *Perpendicularity of Sighting Vanes.*—The sighting vanes should be perpendicular to the base of the ruler. Test by levelling the plane-table, using a level tube if necessary, or set the alidade on a flat surface which is known to be level. Sight a plumb-bob string a short distance away, or the vertical edge of a building, and note if the vertical line so defined appears to be parallel to the hair and sighting vane. If not, put some packing under the base of the vane to tilt it in the direction required.

5. *Adjustment of Indian Clinometer.*—This adjustment consists in making the axis of the bubble tube of the clinometer parallel to the line of sight when the latter is horizontal, and it is made in exactly the same manner as the similar adjustment of the Abney level, say by sighting the same line in opposite directions (p. 118). The adjustment is effected by means of the adjusting screws of the level tube.

The above are the tests for the ordinary plane-table, simple alidade, and Indian clinometer, but if a telescopic alidade is used, additional tests and adjustments of the latter are necessary. These are:

1. *Adjustment of Base-level Tubes.*—These should be parallel to the base of the ruler. Test by bringing one bubble to the centre of its run after the alidade has been placed on the table. Draw a line against the side of the ruler, lift the alidade and reverse it, and then set it with the rule against the same line. If the bubble is no longer central, correct half the apparent error by the level tube adjusting screws and the other half by levelling the board.

2. *Adjustment for Perpendicularity of Line of Collimation and Horizontal Axis of Telescope.*—This involves making the line of collimation perpendicular to the horizontal axis of rotation of the telescope, and is carried out in exactly the same manner as the similar adjustment for the theodolite (p. 102).

3. *Adjustment for Parallelism of Base of Ruler and Horizontal Axis of Telescope.*—This adjustment is similar to the adjustment of the horizontal axis of the transit theodolite (p. 103).

4. *Adjustment of the Telescope Level.*—If there is a level on the telescope, the axis of the level should be made parallel to the line of collimation. Test by the "two-peg" method of testing the Dumpy level (pp.125–128) and use the screws of the level tube for the adjustment.

5. *Adjustment of the Vernier Index.*—The reading on the vertical arc should be zero when the line of collimation is horizontal. This adjustment is done in the same manner as the adjustment of the vertical circle index of the theodolite (pp. 105–108).

5. Setting Up and Orienting the Plane-table: Plane-table Resection.

In ordinary small-scale work with the plane-table very careful and accurate centring and levelling are not essential. If the table is to be set over a definite point on the ground, it is placed centrally over that point, and set up as nearly level as can be judged by eye. If no plumb bob is available, a stone or coin dropped from immediately underneath the centre of the tripod head will show if the instrument is approximately over the ground mark, and a level tube or circular level placed on the table will indicate if the latter is level. Final levelling is done by slight shifts of the tripod legs, or, if a quick-levelling movement, such as that shown in fig. 7.2, is fitted to the tripod head, by loosening the clamping screw under the latter, moving the table until it is level, and then clamping the screw again. During this process, the level should be tried in several different directions, and, when levelling is completed, the bubble should remain central for all directions on the table.

In ordinary work on a small scale it is quite good enough to set the level " by eye " without the use of a level tube.

In large-scale work, and particularly in plane-table traversing with very short legs, it is sometimes necessary to set the table with a particular point on the plan vertically over the corresponding point on the ground, and in that case *a plumbing bar* is necessary. This consists of a hairpin-shaped light metal frame having arms of equal length, in which the plumb bob is suspended from the end of the lower arm. The fitting can be placed with the upper arm lying on top of the table, and the lower arm below it, the table being centred when the plumb bob hangs freely over the ground mark and the pointed end of the upper arm coincides with the equivalent point on the plan.

Orienting the plane-table means putting the board into some fixed direction so that a line representing a certain direction on the plan is parallel to that direction on the ground. Thus, if the direction of the magnetic north is shown on the plan, or all directions on the plan are to be referred to the magnetic meridian, the board should be set so that the line on the plan representing the direction of the magnetic meridian does in fact point in the direction of the magnetic meridian, or so that other lines on the plan point in their true directions relative to the magnetic meridian.

There are two main methods of orienting the plane-table:

1. By means of the trough compass.
2. By means of points fixed on the ground whose positions are accurately plotted on the plan.

In orienting by the trough compass the direction of the magnetic meridian must be shown on the plan by a good firm straight line. Set the side of the box of the trough compass accurately against the line and loosen the clamp below the tripod head which controls the rotation of the table about the vertical axis (the vertical axis clamp). Set the compass needle free to swing and rotate the table until the ends of the needle coincide with the index marks on the scale in the box. Then tighten the vertical axis clamp. The table is now oriented with the line representing the direction of the magnetic meridian actually pointing along that meridian.

In orienting by means of points already fixed on the ground whose positions are plotted on the plan, two cases arise:

PLATE XIII

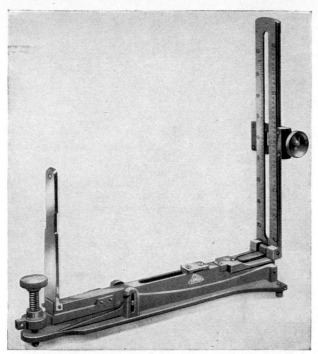

Fig. 7.4.—INDIAN
CLINOMETER
(By courtesy of Messrs
Hilger and Watts, Ltd.

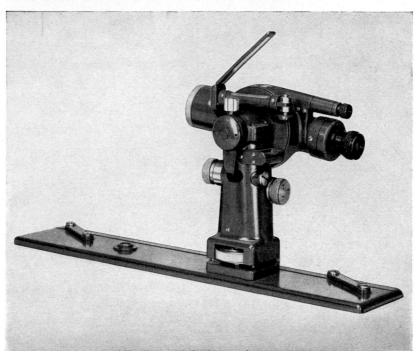

Fig. 7.5.—TELESCOPIC ALIDADE
(By courtesy of Messrs. Hilger and Watts, Ltd.)

a. It is possible and convenient to set the plane-table over one of these points.

b. It is impossible or it is inconvenient to set the plane-table over any of these points.

In the first case, set the instrument over a point whose position is plotted on the plan. Draw a line on the latter between the plotted position of this point, which we shall call the *station point*, and the plotted position of some other point, the *object point*, which can be seen from the station point, and set the alidade so that the edge of the rule coincides with this line. Unclamp the vertical axis clamp and turn the table about its vertical axis until the object point appears to be in line with the sights in the vanes of the alidade. Tighten the vertical axis clamp. The table is then oriented in the proper direction. If possible, check by laying the rule against the line joining the station point and a second object point and seeing if this point appears to be on the line of sight.

If it is not possible or convenient to set the instrument over any of the points plotted on the plan, some method of *resection* must be used. The simplest is to use three fixed points. In this case, the exact position of the point on the plan representing the point where the instrument is set is not known and has to be found, but *it should be noted that a solution is not possible if a circle can be drawn which will pass through the three fixed points and the station point. Consequently, care in the selection of the three fixed points is necessary to see that the station point does not lie on, or too close to, the circle passing through the other three.* Needless to say, all the object points must be visible from the station point.

Select a trial point on the plan which, as far as can be judged, is the proper plotted position of the station. Draw a light pencil line to join this point to the plotted position of one of the fixed points, and lay the edge of the alidade along this line. Then loosen the vertical axis clamp, rotate the table until the chosen fixed point appears to lie on the line of sight of the alidade, and clamp the table in this position. Take a pencil with a very long slender point and hold it vertically with the point on the plotted position of a second fixed point. Still keeping the table clamped and the pencil steady, set the edge of the alidade against the pencil point and, using the latter as a pivot, rotate the alidade until the line of sight appears to be in line with the point whose plotted position is being used as the pivot. Draw a line against the edge of the alidade so that this line passes through the point where the pencil was held. This line, produced if necessary, will meet the first line at a point which, if the point assumed as the station point were in its correct position, would pass through that point. As it is unlikely that the assumed position will be correct, these two points will not coincide. Again, still keeping the table clamped in its original position, repeat with the third fixed point the procedure already used with regard to the second. This will give three lines passing through the plotted positions of the three fixed points, and the intersection of these lines two by two will give a triangle which is called the *triangle of error*. Thus, in fig. 7.7, **a**, **b**, **c** are the plotted

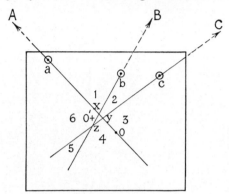

Fig. 7.7.—Three-point resection

positions of the three fixed points A, B and C, **o** is the assumed position of the station point, and **axyo**, **bxz** and **cyz** are the positions of the three lines drawn as above, **x**, **y**, **z** being the intersections of these lines with one another. These intersections give the triangle of error, **xyz**. Scale off on the plan the distances from **o** to **a**, **b** and **c** and let these distances be d_1, d_2 and d_3 respectively. The true position of the point **o** can now be found by the following three simple rules, which should be committed to memory:

I. If the station point is inside the triangle formed by the three fixed points, its proper plotted position falls within the triangle of error. Otherwise, it is outside it.

II. In the latter case, the plotted position will be such that it is either to the right or to the left of all the rays as the surveyor faces the fixed points. There are only two sectors in which this condition can be fulfilled.

III. The exact position of the point can now be found by the condition that the distances of the plotted point from the three rays are proportional to the lengths, d_1, d_2 and d_3, of these rays to the fixed points.

Applying these rules to fig. 7.7, we see that by Rule I the point lies outside the triangle **xyz**; by Rule II it lies in either of the sectors 3 or 6, and, by Rule III, if d_1, d_2, d_3 are proportional to 3, 2 and 3, say, the proper position is at **o′**, where **o′** is at distances proportional to 3, 2 and 3 from **axy**, **bxz** and **cyz** respectively.

In fig. 7.8 the point lies inside the triangle formed by the three fixed points, and hence, by Rule I, the plotted position lies inside the triangle of error **xyz**, and, by Rule III, it must be at **o′**, where the distances from **o′** to the lines **ayx**, **bxz** and **czy** are in the ratio $3 : 2 : 3$.

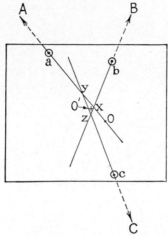

Fig. 7.8.—Three-point resection

Having found **o′**, set the alidade with the edge along one of the rays from **o′** to **a**, **b** or **c** and unclamp the table. Revolve it slightly about the vertical axis until the line of sight through the vanes intersects the fixed point corresponding to the chosen ray, and then clamp the table with the alidade pointing in this direction. The plane-table should now be properly oriented, and the point **o′** on the plan should be the true plotted position of the point where the table is standing. This should be verified by laying the alidade along the rays **o′b** and **o′c**, or along a ray **o′f** to a fourth fixed point F, and seeing if the line of sight intersects the corresponding points on the ground. If not, there is an error of some sort, and the whole of the previous procedure must be repeated.

It is also possible to orient a plane-table—and to obtain a fixing at the same time—by the use of two fixed points only, but this involves the use of an auxiliary station point. Let A and B be the two fixed points, and let **a** and **b**, fig. 7.9, be their plotted positions on the plan. Let O be the point where the instrument is to be set up and oriented. Choose an auxiliary point O′ from which A, B and O can be seen and which, with O, will give well-shaped triangles O′OA and O′OB. Set the instrument at O′ and choose the point **o** on the plan as the approximate position of O′. Orient and clamp the table so that when the alidade is set along the line **o′a** the line of sight intersects the point A. Then sight the points B and O with the alidade edge kept

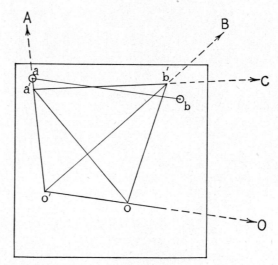

Fig. 7.9.—Two-point resection

against **o′** and draw the lines **o′b′** and **o′o** in the directions of B and O respectively. Move the instrument to O and set up there. Estimate or pace the distance O′O and plot the point **o** on the line **o′O** as the approximate position of O. Set the alidade along **oo′** and orient and clamp the table with the line of sight of the alidade directed to the point O′. Pivoting the alidade about **o**, draw lines through **o** in the directions OA and OB to cut **o′A** in **a′** and **o′B** in **b′**. As the table is not yet properly oriented, and the scale of **o′o** is not quite correct, the points **a′** and **b′** will not coincide with **a** and **b**, but, by the principle of triangulation, the direction of **a′b′** will be the true direction of AB, our object being to bring **ab** into this direction. Hence, to orient the table, set the alidade along **a′b′**, and set out a pole some distance away at a point C so that the pole is seen on the line of sight of the alidade.

Unclamp the table, set the alidade on the line **ab**, and turn the table until the line of sight intersects the pole at C and then clamp. The table will now be properly oriented at station O, and, if the alidade is set against **a** and **b** and lines drawn in the directions of A and B, these lines will intersect at a point which is the true plotted position of the point O.

The process just described is called *two-point resection*. The first process, using three fixed points, is called *three-point resection*, or, more usually, simply *resection*. It is to be noted that each process gives both an orientation and a fixing.

CHAPTER VIII

PHOTOGRAPHIC SURVEYING

1. Methods of Photographic Surveying and their Application.

There are two kinds of photographic surveying, the one from the ground and the other from the air. The former is what is usually meant by the term *photographic surveying*, the latter being called *air photo-surveying*, or simply *air survey*. Air survey is a highly specialized subject, with its own technique, so that in the following pages we shall concern ourselves more particularly with the principal apparatus used in ordinary ground photographic surveying—the photo-theodolite—and be content with only a brief description of a modern air survey camera.

Ground photographic surveying is not very extensively employed either here or in the Empire, but it is popular in parts of the Continent, especially in mountainous countries such as Switzerland. Its main advantage is for work in very mountainous country, and in such country it may be successfully used where any other form of survey, other than air survey, would be virtually impossible. For this reason, it has been extensively used in Canada in connexion with surveys in the Rockies. In the end, it is a graphical method of survey, and hence, in ordinary country, not so accurate as work done with the theodolite and steel band. Neither is it so accurate as ordinary plane-tabling, the main reason being that the survey must be plotted, usually on a larger scale than the photographs, from material which suffers distortion during the processes through which it has to go before it is used. Thus, there is some distortion caused by imperfect correction of the camera lens (almost negligible with a good modern lens), and there is further distortion during the development and drying of the negative material, film or coated glass, used in the camera, and still more in the processing of the final prints. Because of this distortion, plates are used in ground survey cameras instead of film, but in air survey photography the negative material is usually film.

PLATE XIV

Fig. 8.1.—PHOTO-THEODOLITE

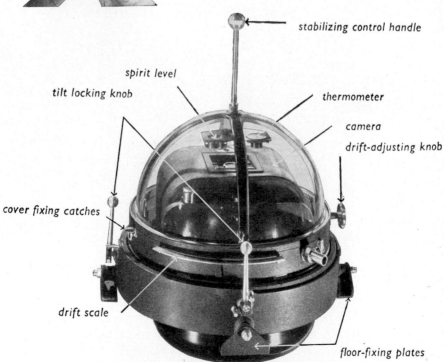

Fig. 8.4.—WILLIAMSON O.S.C. Mk. I AIR SURVEY CAMERA
(By courtesy of Messrs. The Williamson Manufacturing Co., Ltd.)

THE PHOTO-THEODOLITE

2. Description.

The photo-theodolite, fig. 8.1, is merely a theodolite in which the upper plate and upper part of the instrument are replaced by a plate camera. On top of the camera is a telescope and vertical graduated arc, and below it is a vertical central spindle to take the place of the vertical axis of the upper circle. As it is never necessary to take photographs of very near objects, the camera consists of a solidly constructed metal or wooden box with a fixed-focus lens in front and a slide in the back in which a plate holder, containing the plate, can be inserted. Inside the box, and near the back, is an open metal frame carrying a vertical and a horizontal wire, these wires lying in the vertical and horizontal planes passing through the optical axis of the lens. When the sheath of the plate holder is drawn, the frame can be racked back by means of a milled head outside and at the side of the box to touch the sensitized surface of the photographic plate, thus causing the two wires to register very fine vertical and horizontal lines on the plate when the latter is exposed. In the Bridges-Lee photo-theodolite shown in fig. 8.1 the bottom of the frame also carries a transparent graduated scale, the graduations showing to the nearest 5′ the angular distances from the central wire. Consequently, an image of the scale, in which the numbers and graduations are marked in dense black, appears at the top of the negative, and angular distances from the central wire may be read direct on the print. This particular instrument is also provided with a magnetic compass in the bottom of the camera box, the needle of which carries a vertical transparent scale, marked off in degrees and half-degrees. This scale registers on the negative so that magnetic bearings may be read off direct on the print if desired. The compass needle is released to swing on its pivot when the frame carrying the cross hairs is racked back.

The lower part of the instrument consists of the lower part of a theodolite, including graduated lower plate and levelling head, and a vernier attached to the back of the camera enables readings on the graduated arc of the lower plate to be taken to single minutes of arc.

The telescope and vertical arc on top of the camera are mounted in such a way that the vertical hair in the telescope and the vertical wire in the camera both lie in the vertical plane through the optical axis of the camera lens. A level tube is mounted on the telescope, and a similar tube on top of the body of the camera serves to indicate when the vertical axis is truly vertical.

In some cameras the horizontal and vertical wires in the frame are omitted, their place being taken by notches or pointers which show as such at the top and bottom and sides of the picture. Lines drawn on the photographs to connect opposite notches or pointers then replace the marks of the ordinary cross wires.

It will be seen that, although it is a great convenience to have both camera and theodolite combined in one instrument, this is not absolutely necessary, and a separate camera and theodolite would serve instead. In this case, the camera must be provided with levels to ensure that the plane of the plate is vertical at the moment of exposure, and also with means of registering on the photograph the positions of the vertical principal line and the horizontal line.

It is very important in all cameras intended for surveying that the lens should give a picture free from appreciable distortion, but, as the camera of a photo-theodolite is always used on a stand and the pictures to be taken are ordinary landscapes, there is no need for the lens to be specially fast—one of about $f/6\cdot3$ is usually all that is required.

3. Determination of Focal Length of Camera Lens.

It is necessary to know the focal length of the lens, and, if this length is not already known, it can best be found by a method due to the late Dr. Deville, at one time Surveyor-General of Canada, and one of the pioneers of photographic surveying in the Empire.

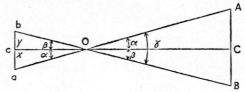

Fig. 8.2.—Determination of the focal length of the camera lens

Having set up and levelled the instrument, select two *distant* points A and B, shown in place in fig. 8.2, both of which will show on the photograph and come fairly close to the edges of the field, and measure the angle between them by using the instrument as a theodolite. Call this angle γ. Now, using the instrument as a camera, take a photograph which will include both points. Let **a** and **b** on the plan be the images of these points on the plate and **c** the mark of the vertical cross wire. On the plate (not on the print), measure the horizontal distances **ca** and **cb**, and call these distances x and y. Since the points A and B are very distant, the distance **c**O is equal to f, the focal length of the lens. Hence, from the figure,

$$\tan \alpha = \frac{x}{f}, \tan \beta = \frac{y}{f}.$$

$$\tan \gamma = \tan (\alpha + \beta)$$

$$= \frac{\tan \alpha + \tan \beta}{1 - \tan \alpha \tan \beta}$$

$$= \frac{\dfrac{(x + y)}{f}}{\left(1 - \dfrac{xy}{f^2}\right)}.$$

Hence,

$$f^2 \tan \gamma - xy \tan \gamma = (x + y)f,$$

or,

$$f^2 \tan \gamma - (x + y)f - xy \tan \gamma = 0;$$

whence

$$f = \frac{(x + y)}{2 \tan \gamma} + \sqrt{\frac{(x + y)^2}{4 \tan^2 \gamma} + xy},$$

the positive sign being taken for the second term because this term is numerically greater than the first, and f is to be taken as a positive quantity.

4. Determination of Principal and Horizon Lines.

The *principal* and *horizon* lines of the camera are the vertical and horizontal lines where the vertical and horizontal planes through the optical axis of the lens meet the plane of the plate, and the point of intersection of these lines, i.e. the point where the optical axis meets the plate, is the *principal point* of the camera. The principal and horizon lines should coincide with the vertical and horizontal lines traced by the cross wires of the camera on the photograph. To test this relationship proceed as follows:

After levelling carefully, take a photograph of a suspended plumb bob and see if the string of the plumb bob, as it appears in the negative, is parallel to the trace of the vertical wire in the frame. Then see if the trace of the horizontal hair is at right angles to the trace of the vertical hair.

Assuming the tests for verticality and horizontality to be satisfied, set up and level the instrument at any convenient point and select three distant points A, B and C, such that the images of all three points will fall in a single photograph. Measure the horizontal angles between each pair of points, and the vertical angle to one of them, say A, and then take a photograph to include all three points. On a plain sheet of paper draw in the three rays OA, OB and OC converging from the point O representing the point of observation and making

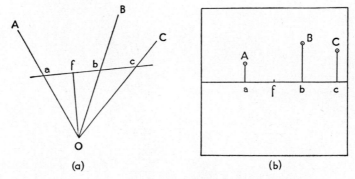

Fig. 8.3.—Determination of the principal and horizon lines

angles with each other equal to the observed angles, as in fig. 8.3a. Through the images A, B and C of the three points on the photograph draw verticals to meet the trace of the horizon wire at **a**, **b** and **c**, as in fig. 8.3b. Lay a sheet of paper with a straight edge against this line, and mark off on the paper sheet the positions, **a**, **b** and **c**, of the points on the line where it is met by the verticals through the images of the points. Take this sheet of paper and move it about on the plot of the rays until each of the three points on the edge falls on its corresponding ray, as in fig. 8.3a. Let **abc** be the position of the paper so found. Draw in the line **abc**, and from O draw a line O**f** perpendicular to it. Then **f** is the position on the paper where the principal line meets the trace of the horizontal wire, and O**f** is equal to the focal length of the lens. Lay the sheet edge, with this point marked on it, along the trace of the horizontal wire on the photograph with the

points **a**, **b** and **c** on it lying against the corresponding points on the
photograph. Then the point where the mark at **f** intersects the trace
of the horizontal wire is the point where the principal line intersects
this wire. Consequently, a vertical line drawn through **f** (i.e. a line
drawn through **f** parallel to the vertical wire) defines the principal line.

To verify the position of the horizon line, let A be the point to
which a vertical angle has been observed by theodolite and let this
angle be α. Measure O**a** in fig. 8.3a. Then the vertical distance from A
on the photograph to the horizon line is O**a** tan α. Scale off this distance
on the photograph vertically below the image of A. This gives a point
on the horizon line, and a line drawn through this point perpendicular
to the principal line will be the horizon line, the intersection of the
principal and horizon lines being the principal point.

5. Using the Photo-theodolite.

Set up and level the instrument and direct it at the view to be
taken, the view-finder or ground-glass screen being used to judge
the exact limits of the picture. After having inserted the plate holder
and drawn the slide, rack back the frame carrying the cross hairs
to touch the plate. The exposure to be given is best determined by
means of an ordinary photographic exposure meter. Set the diaphragm
to the proper stop and the shutter to the appropriate speed and make
the exposure. Rack the cross-hair frame forward to clear the plate,
close the slide, and withdraw the plate holder to make room for another
holder or for the camera back to be replaced if no more photographs
are to be taken at the station.

The sensitive material used in the photo-theodolite preferably consists of coated glass plates, as ordinary film is much more liable to
uncontrollable distortion in the developing, fixing, and washing processes. Good, sharp, clear pictures, full of detail, are required. Consequently, it is well to use a small stop and a correspondingly relatively
long exposure. The best emulsion is orthochromatic or panchromatic
used in conjunction with a suitable colour filter on the camera.

As each plate is exposed, notes should be made in the field, giving
particulars of station of observation, description of views taken, list
of permanent points included in the picture, date, time, weather, stop
and length of exposure.

AIR SURVEY CAMERAS

6. Methods of Air Survey and the Single-lens Camera.

Cameras for air survey work can be divided into two classes: (1) single lens and (2) multiple lens. The single-lens camera is designed to take " vertical ", or almost vertical, photographs, and is the type ordinarily used by the R.A.F. as being most suited for normal accurate work. In the multiple-lens model, a number of lenses are fitted, of which one is set vertically and the remainder obliquely with reference to the horizontal camera base, so that one exposure gives a set of one vertical and two or more oblique photographs equally oriented about a vertical axis. This type of camera has been extensively used for the air survey of some of the undeveloped areas of Canada, the resulting maps being plotted on a very small scale. The multiple-lens camera, in fact, is best suited for rapid surveys on a small scale of fairly flat undeveloped areas, the single vertical lens being better adapted for deliberate mapping on large and intermediate scales. Hence, as the latter is the type of work which is of most interest to engineers, we shall content ourselves here with a very brief description of the single-lens camera.

Since a camera taking vertical photographs should be as free as possible from vibrations and other uncontrollable movements of the aircraft, it is usually mounted on gimbals carried on a vibrationless support. The lens is of the wide-angle type, so as to cover a large area with a single exposure. Means are provided, in the form of a pressure or a suction plate, for keeping the film completely flat and in register with the collimating marks when it is being exposed, and a suitable shutter of the rotary disc or louvre type is fitted. (The focal-plane shutter is not suitable for vertical air survey.) Moreover, as most flying is done at a fairly considerable height where temperatures may be very low, it is desirable to provide some system of heating the apparatus to prevent frosting of the lenses, sluggishness and sticking of moving parts, and cracking of the film. Maintaining a steady temperature also has the effect of reducing distortions due to contraction and expansion of optically calibrated parts and other components of the camera. In addition, the camera includes an arrangement by which tilt, altitude, time, number of exposure, and other data are photographically recorded on the side of each photograph.

7. Air Survey Camera O.S.C. Mk. I.

Fig. 8.4 (Plate XIV, p. 179) shows the Air Survey Camera O.S.C. Mk. I recently produced by The Williamson Manufacturing Co., Ltd., of Willesden Green, London, N.W.10. This camera accommodates 430 ft. of $9\frac{1}{2}$-in. film, providing 500 exposures, each 9 in. square, plus instrument records adjoining each exposure. The transparent-dome arrangement on top is part of the patented heated enclosure in which the camera proper operates, so that a suitable steady temperature may be maintained under varying atmospheric conditions. The heat for this installation is derived from the ordinary cabin heating supply. The normal optical equipment consists of a 6-in., $f/5.5$ Ross wide-angle lens, but $8\frac{1}{4}$-in., 12-in. and 25-in. lenses and body units can be had if required, the 6-in. lens giving a field of view of some 96°. A between-lens sector shutter is provided giving speeds up to 1/300 of a second. The shutter is operated by remote control to give accurate exposure intervals over a range of from 2 to 60 seconds, remote control of the shutter speeds also being fitted. The operation of the camera is all-electric, arranged in such a way that the winding mechanism for the film is electrically interlocked with shutter and instrument exposure so that all may work together in proper sequence.

The complete camera enclosure is supported on two horizontal tilting axes, set at right angles to one another, each of which can be separately locked by special handles outside the casing. In addition to acting as locks, these handles can be used to provide the necessary amount of friction to enable operations by the stabilizing control handle to be carried out without excessive effort and without shifting the camera. The stabilizing control handle, seen at the top of the cover, enables the camera to be controlled, if desired, in accordance with the indications of a spirit level on top of the camera immediately under the transparent cover.

CHAPTER IX

SUPPLEMENTARY NOTES

SELF-ALIGNING LEVELS

1. Self-aligning Levels.

In recent years, different makers have produced levels in which it is only necessary to level the instrument very approximately, the final setting of a horizontal line of sight being affected by gravity acting on a pendulum or a freely suspended reflecting surface. The simplest of these instruments is the Cowley level intended for builders, architects, &c., but a much more elaborate and accurate instrument, intended for every class of work, has been manufactured for some years by the German firm of Zeiss, while more recently, in this country, Messrs Cooke, Troughton & Simms have collaborated with Messrs Hilger & Watts in bringing out a common system of control—a corrector or stabilizer—

Fig. 9.1.—Cowley Automatic Level in use

(By courtesy of the makers)

which each firm embodies in a level that otherwise retains the features which characterize the other levels made by it.

The Cowley Automatic Level, which is manufactured by Messrs Hilger & Watts, Ltd., is best suited for giving levels on building and engineering works of all kinds rather than for carrying through long lines of levels, but it has the advantages of being cheap, small ($4\frac{3}{4}$ × $5\frac{1}{2}$ × 2 inches), light in weight ($2\frac{3}{4}$ lb. without tripod), quick to set up, and not needing a skilled surveyor to use it. In appearance it looks more like a small cine camera than a level. It is used in conjunction with a vertical graduated staff carrying a horizontal target, 18 inches wide by 2 inches deep and with a heavy black sighting line in the middle, which can be moved up and down the staff in accordance with signals or directions given by the surveyor. When the instrument is set on its tripod on a spike designed to receive it, a pendulum and mirror system is

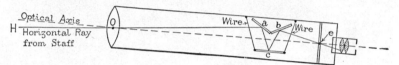

Fig. 9.2.—Diagram to illustrate method of operation of Zeiss Self-aligning Level. Ni.2

automatically released to give a horizontal line of sight. When the staff is viewed through an aperture on top of the casing, the two ends of the target will appear to be separated, one above the other, if the thick sighting line on the target is not exactly on a horizontal line passing through the optical centre of the instrument. If the thick sighting line on the target is exactly on this horizontal line, its two ends will appear to coincide, so that the image represents a continuous straight horizontal line. Thus, the surveyor does not read the staff himself but directs the staff man to move the target up or down until coincidence of the two images is obtained, the staff man singing out the reading on the graduated side of the staff which faces him.

The Self-Aligning Level Ni.2 of Messrs Zeiss, and the more recent ones of Messrs Cooke, Troughton & Simms and Messrs Hilger & Watts, in appearance resemble an ordinary conventional level and are intended for much more accurate and general work than the Cowley level. In fact, they are capable of doing all the work generally expected from a large level of the conventional type.

In the Zeiss level, two mirrors, a and b in fig. 9.2, are permanently fixed to the barrel of the telescope and a mirror c is suspended by four fine wires from the top of the barrel so that it tilts when the telescope is tilted. This system of two fixed mirrors and one swinging one is called the *compensator*, and its dimensions and its position in the telescope are so designed that, when the telescope is tilted through any small angle, a horizontal ray entering it through the optical centre of the objective is reflected, as shown by the full line HOacbe in fig. 9.2, so as always to coincide with the horizontal wire e of the diaphragm. Thus, when the observer sees the image of an object coinciding with the cross hair, that object must lie in a horizontal plane through the optical centre of the objective. In practice, in order to reduce loss of light to a minimum, the mirrors are replaced by reflecting prisms.

The wires supporting the horizontal reflecting surface are of non-magnetic material and are very strong for their weight and thickness, and the compensator is fitted with a special damping device which not only damps down the oscillations of the mirror and brings it to rest very quickly but also reduces the effects of ground vibrations to a mimimum, so that the instrument is remarkably free from disturbance by vibrations from heavy traffic. Temperature changes have little or no effect and a parallel-plate micrometer (page 134) can be obtained as an extra, which, attached to the instrument, turns the latter into a precise level.

The system used in the Cooke, Troughton & Simms and Hilger & Watts self-aligning levels, though differing in details of design, produces the same result as the compensator in the Zeiss instrument; that is, for all small tilts of the telescope, a horizontal ray passing through the optical centre of the objective is made to pass through the horizontal cross hair in the diaphragm. In this system, however, the middle reflecting surface is rigidly attached to the top of the barrel, and the two outer reflecting surfaces are combined in a swinging gravity-controlled mount so that they move as a whole when the telescope is tilted but remain in their original position relative to the horizontal. The mount is not supported on wires but on an ingenious flexure pivot formed by four crossing flexible blades or strips of metal which connect the movable and stationary parts, the point of crossing of the blades being the point of rotation. This form of pivot is virtually frictionless and at the same time is very robust. A damping arrangement ensures that the system comes to rest in a fraction of a second.

WEDGE TELEMETERS AND SELF-REDUCING SPLIT-IMAGE TACHEOMETERS

2. Wedge Telemeters.

The accuracy of a length measured with the ordinary tacheometer with stadia hairs is seldom much better than 1/500, which for much traverse work is not accurate enough. To meet the demand for one capable of more accurate work, special types of instruments have been developed on the Continent in recent years which enable short lengths of up to about 500 ft. to be measured with an accuracy of anything

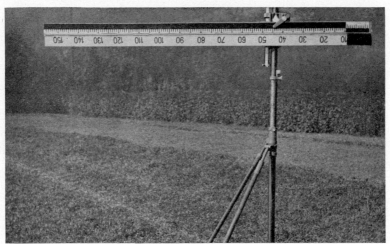

Fig. 9.3.—Horizontal staff used in conjunction with the Wild Precision Telemeter DMI.

(By courtesy of the makers)

between 1/3000 and 1/10,000. They are all used with a horizontal staff supported on a graduated pole provided with two stay rods (fig. 9.3). Running along the face of the staff is a horizontal line below which is a scale graduated in centimetres, and above it, and at the zero end, is a short vernier scale. When this staff is perpendicular to the direction of the instrument and is viewed in the latter, the graduated scale is seen above the line (fig. 9.4), and below the line is seen an image of the vernier displaced along the scale by an amount which depends on the distance of staff from instrument. The reading on the staff short of the zero of the vernier gives the number of metres in the required

8*

distance. Then, by turning a graduated drum on the instrument, the operator can bring a graduation of the vernier into exact coincidence with a graduation on the staff, and the reading on the vernier will give the number of decimetres in the distance, while the reading on the drum

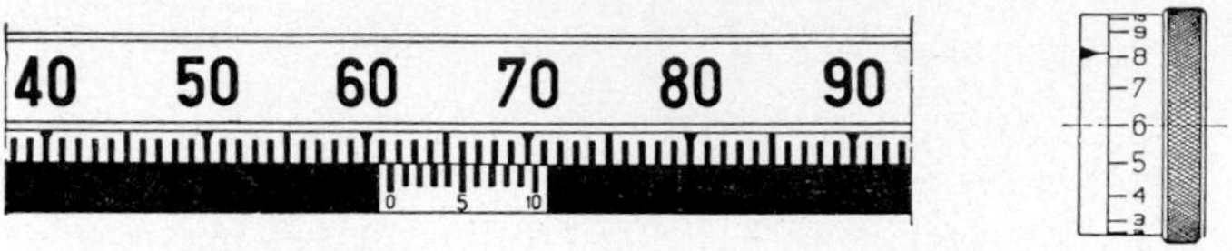

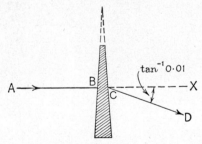

Reading on staff	-	-	-	-	-	61	m.
Vernier	-	-	-	-	-	-	0·5	m.
Reading on drum	-	-	-	-	-	0·08	m.
Total reading	-	-	-	-	-	61·58	m.

Fig. 9.4.—Example of reading

(By courtesy of the makers)

will give the number of centimetres in the distance. By estimating tenths on the graduations on the drum the distance is given (in theory) to the nearest millimetre.

3. Principle of The Wedge Distance Telemeter.

The wedge distance telemeter depends for its operation on the fact that, on passing through a wedge or prism-shaped piece of glass, a ray of light (AB in fig. 95) is diverted in direction so that on emergence it

Fig. 9.5.—Deviation of ray of light by glass wedge

makes an angle XCD with its original direction. The amount of divergence depends on the angle of the wedge and the refractive index of the glass, but in practice the wedge is designed so that the angle of divergence is $\tan^{-1} 0·01$ and the distance to the staff is thus 100 times the deviation produced on it.

The wedge telemeter manufactured in this country by Messrs Hilger & Watts, and by Messrs. Wild Heerbrugg and Messrs Kern on the

Continent, consists of a detachable wedge or prism unit which can be fitted in front of the objective of certain of their theodolites. This attachment, which is balanced by a detachable counterweight suitably placed at the eyepiece end of the telescope, consists of a fixed achromatic glass prism or wedge and a plano-parallel glass plate rotatable about a vertical axis by means of a micrometer screw with graduated drum, this plate and micrometer screw forming a parallel-plate micrometer as described on page 134. The wedge, which, with the parallel glass plate, covers the middle third of the theodolite objective, is ground

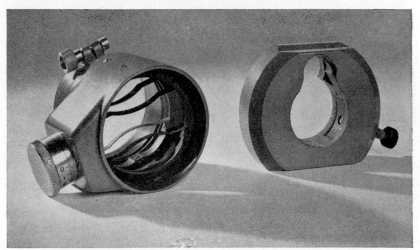

Fig. 9.6.—Precision Wedge Telemeter. Wild DMI, with Counterweight
(By courtesy of the makers)

so that a ray of light entering it is deviated through an angle of $\tan^{-1}$ 0·01. When the instrument is directed to view the staff, rays from the latter passing through the part of the lens not covered by the wedge and plate are not deviated, but those passing through the wedge are deviated so that the observer sees a normal image of the staff on which is superimposed an image of the vernier deviated through an angle of $\tan^{-1} 0·01$. Consequently, as explained above, when the micrometer screw is turned to bring a graduation of the vernier into exact coincidence with the appropriate centimetre division on the staff, the staff, vernier, and micrometer readings when added together will give the distance in metres, decimetres and centimetres.

With careful work, the accuracy of such an attachment may be anything from 1/3000 to 1/5000, or even greater. The range is from 20 to

150 metres in the case of the Hilger & Watts attachment, and 10 to 150 metres in the case of the Wild instrument. A special sighting device on the staff support enables the staff to be aligned perpendicular to the line of sight, and a spirit level serves to indicate when it is horizontal.

4. The Split-image Self-reducing Tacheometer.

The wedge telemeter gives the inclined distance of staff to instrument when the two are not at the same level, so that this distance has to be reduced subsequently to its horizontal equivalent. In the split-image self-reducing tacheometer RDH of Messrs Wild Heerbrugg, the staff reading multiplied by the instrumental constant of 100 gives directly the reduced horizontal distance, and another reading the reduced difference

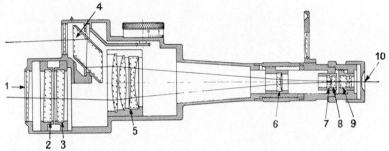

Fig. 9.7.—Diagram to illustrate the principle of the Wild RDH Split-image Self-reading Tacheometer

(By courtesy of the makers)

of elevation between instrument and staff; while the wedge telemeter superimposes an image of the vernier end of the staff on one of the other end, the split-image tacheometer forms superimposed images of the staff and vernier one above the other but with unwanted parts of the images eliminated. Thus, a sharper image of staff and vernier is obtained.

The principle of the instrument will be understood from fig. 9.7 which shows diagrammatically a vertical cross-section of it. The rhomboid reflecting prism 4 on top admits rays which enter the upper part of the objective 5 and form an undeviated image of the staff. The adjustable glass cover, 1 is slightly wedge-shaped and 2 and 3 are achromatic glass wedges which, as the telescope is raised or lowered, rotate in opposite directions an amount proportional to the angle of elevation or depression. The glass cover 1 is used for a final fine adjustment to make the angle of deviation of the ray on emergence from the wedge 3 exactly equal to $\tan^{-1} 0\cdot01$ when the instrument is set for zero elevation or depression. Number 6 is an internal focusing lens and 7 a glass prism set in the focal plane of the objective which, with the slit diaphragm 10, serves to separate the superimposed images.

In the diagram, the instrument is set for the measurement of differ-ences of elevation with the optical axis horizontal, and the wedges so placed relative to one another as to produce zero deflection. As the telescope is rotated, the wedges are rotated in opposite directions by means of a train of gears to produce horizontal deflections proportional to sin β, the factor by which the inclined distance must be multiplied to give the difference in elevation for an angle of elevation or depression, β; the vertical deviations caused by the individual wedges cancel out.

When horizontal distances are to be measured, the rotatable wedges are turned as a unit through 90° by means of a knob on one standard to

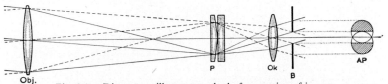

Fig. 9.8.—Diagram to illustrate method of separation of images
(By courtesy of the makers)

produce a combined maximum deflection when the optical axis is hori-zontal. Rotation of the telescope then causes the wedges to rotate in opposite directions to produce a horizontal deflection proportional to cos β, the factor by which the inclined distance must be multiplied to give the true horizontal distance.

Fig. 9.8 shows the method of superimposing the images of staff gradu-ations and vernier. P is prism 7 in fig. 9.7, the middle edge of which forms a horizontal line which is set during measuring operations to coincide with the middle of the horizontal staff and which forms a line of separation between the two images. Rays coming from the rhomboid prism 4 in fig. 9.7, the continuous lines in fig. 9.8, fall on the prism P and are deviated there, those coming from the upper half of the objective passing through the eyepiece Ok and the slot diaphragm B, while any coming from the lower half of the lens are stopped by the diaphragm. Similarly, rays coming from the wedges 2 and 3 in fig. 9.7 which fall on the lower part of the objective pass through the diaphragm, any falling on the upper half of the objective being stopped by the diaphragm. In the telescope, therefore two semicircular exit pupils, AP in fig. 9.8, are formed, bisected by the horizontal middle edge of the prism P; when this line is set to coincide with the central line of the staff, the image of the main graduations, which comes via the wedges in front of the objective and the upper part of the prism P, appears above the line, and the image of the vernier, which comes from the rhomboid prism in front of the objective and the lower half of the prism P, appears below the line; the prism brings the exit pupils, which would otherwise merely touch one another, together, and the slot diaphragm cuts out unwanted parts of them.

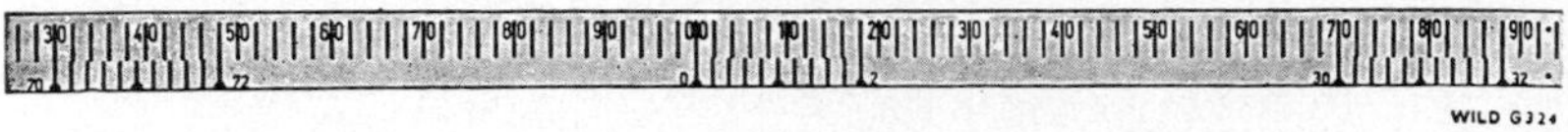

Fig. 9.9.—View of staff as seen in telescope of Wild RDH tacheometer
(By courtesy of the makers)

When it is viewed in the telescope, the horizontal staff used with the tacheometer (fig. 9.9), is seen to have one set of graduations running from 0 to 90 on the right of the zero and another set to the left of the zero but numbered 90, 80, 70, . . . 30, the graduation numbers in this case being coloured red while the figures to the right of the zero are

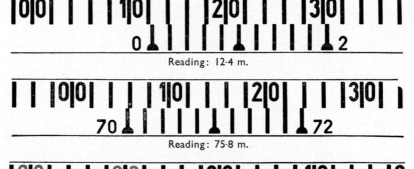

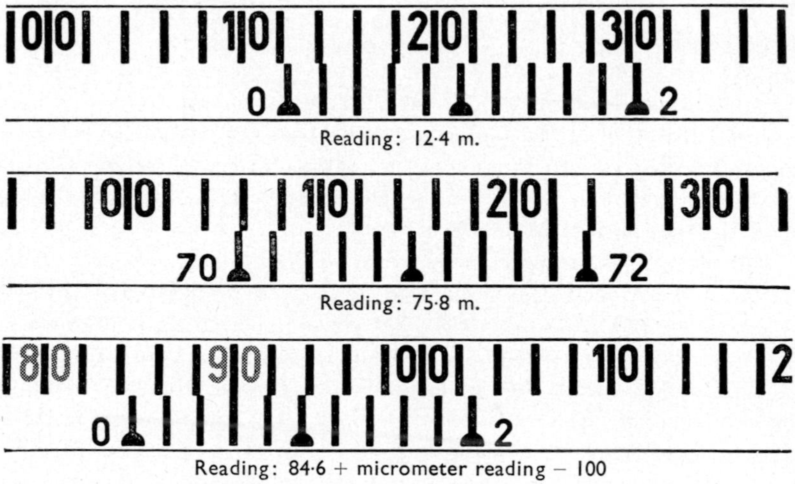

Reading: 12·4 m.

Reading: 75·8 m.

Reading: 84·6 + micrometer reading − 100

Fig. 9.10.—Examples of staff readings
(By courtesy of the makers)

coloured black. There are also three verniers, one with black numbers 0—2 in the centre, one with black numbers 70—72 on the left and one with red graduation numbers 30 and 32 on the right. The central vernier is used for all distances from 0 to 70 metres, but, when the distance of instrument to staff exceeds 70 metres, the central vernier runs off the scale and the vernier on the left must be used, 70 being added to all readings. When angles of elevation are being determined, the black verniers and black numbered scale are used and readings taken in the ordinary way give the reduced difference in elevation. When, however, angles of depression are being observed, the left part of the scale with red graduation numbers is used and the scale reading subtracted from 100. If the difference in elevation exceeds 70 metres, the black vernier runs off the scale and the vernier with red numbers 0 and 2 must then be used, the reading being subtracted from 200.

The staff is graduated at 2-cm. intervals corresponding to 2-m. in distance; the vernier gives the 0·2-m. values and the micrometer screw the 0·01-m. values. Examples of staff readings are given in fig. 9.10.

The instrument can, when required, be employed as an ordinary theodolite by turning a knurled sleeve on the eyepiece of the telescope to the right until it reaches a stop. This brings into the field only the non-deflected image formed by rays entering the rhomboid prism. Readings on both circles are then by micrometer scale reading direct to single minutes.

With reasonably careful work, an accuracy of between 1/5000 and 1/10,000 may be obtained with this instrument.

DISTANCE MEASUREMENT BY ELECTRONIC MEANS

5. Electronic Distance Measurement.

On pages 31-36 we have described the apparatus used in the measurement of a trigonometrical base with Invar tapes or wires and the details and theory of base measurement are described in pages 74-88 of " Principles of Surveying." Such a measurement is somewhat laborious, takes time, and, in general, involves a good deal of clearing and preparation of the line to be followed. During the last ten years, however, two instruments have been devised with which lines between 15 and 30 miles can be measured with an accuracy of about 1/300,000 to 1/500,000, this being not far short of the accuracy obtainable with Invar tapes. The apparatus is expensive but normally it is much quicker to use than Invar tapes and demands practically no preparation of the ground, or laying out and clearing of the line, such as is needed in base measurement with Invar tapes. In fact, these instruments, of which there are already a number in use in different parts of the world, will probably replace Invar tapes and wires in all extensive triangulation work in future.

The instruments in question depend for their operation on measuring the time taken for a signal, or pulse, of light or radio waves sent from the instrument to travel to a distant point where there is another instrument that receives and transmits the signal back to the sending instrument. If this time is known, the distance can be calculated from the velocity of transmission of the signal, supposed known. The velocity of light or electromagnetic waves *in vacuo* (299792·5 km. per sec.) is now known with great accuracy, but this value has to be corrected for air temperature, pressure, and humidity, to get the velocity in free air at the time of measurement. Observations of these quantities have therefore to be taken at both ends of the line and the chief source of error in the operation is the difficulty of getting their true values along the path of the signal as against the values recorded at the ends.

Two separate instruments of the kind are available. The first is the geodimeter developed by Dr. E. Bergstrand in Sweden and the second is the tellurometer developed by Dr. T. L. Wadley in South Africa. The geodimeter operates with ordinary light waves, but the tellurometer operates with radio waves of length 10 cm. (3000 megacycles per second which penetrate haze and mist much better than light waves do and thus enable the tellurometer to be used when work with the geodimeter would be impossible.

The ease with which the geodimeter and tellurometer can be used makes possible in some cases the replacement of triangulation by *trilateration*, i.e. a system in which the lengths of the sides of triangles, and not the angles, are measured. It also makes it possible to introduce more check bases in triangulation than have been customary up to now.

Ordinary radar is another form of electronic distance measurement which can be applied in surveying. Its chief use is in the measurement of very long lines, from 50 to about 400 miles, to give an accuracy of something between 1/20,000 and 1/75,000, so that it can sometimes be employed to bridge large water gaps which are too wide to permit the use of ordinary triangulation. It involves the use of an aeroplane which flies across the line somewhere near its middle and carries a combined self-recording sending and receiving apparatus that instantaneously records the distances from it to two " reflectors " situated at either end of the line. A number of observations are taken as the aircraft crosses the line, and the length of the latter is the minimum sum of the two lengths reduced to ordinary ground level.

Shoran, a system of radar used by the Canadian Geodetic Survey, has been employed extensively to provide, by means of a trilateration with very long sides, a widely spaced control for air mapping on very small scales.

APPENDIX

Proof of Lapse Rate Formula for Relation Between Atmospheric Pressure and Elevation of Station

The mass of a small layer of air of unit area, density ρ, and height dh is $\rho\,dh$, and the increase in pressure with increase in elevation is consequently

$$dp = -g\rho\,dh,$$

the negative sign being used because pressure decreases with increase in height.

From the gas equation,

$$p = c\rho T$$
$$= c\rho(T_e - kh).$$

Hence,

$$dp = -\frac{gp}{c(T_e - kh)}\,dh.$$

$$\therefore \frac{dp}{p} = -\frac{g}{c(T_e - kh)}\,dh.$$

Whence, by simple integration,

$$\log p = \frac{g}{ck}\log(T_e - kh) + \log A,$$

where A is a constant of integration.

$$\therefore p = A(T_e - kh)^{\frac{g}{ck}}.$$

But
$$p = p_e \text{ when } h = 0.$$

$$\therefore A = \frac{p_e}{(T_e)^{\frac{g}{ck}}}$$

and
$$p = p_e\left(\frac{T_e - kh}{T_e}\right)^{\frac{g}{ck}}.$$

INDEX